# Robotics: Design, Construction and Applications

# Robotics: Design, Construction and Applications

**Allie Weaver**

**C WILLFORD PRESS**

www.willfordpress.com

Published by Willford Press,
118-35 Queens Blvd., Suite 400,
Forest Hills, NY 11375, USA

ISBN: 978-1-64728-337-7

**Cataloging-in-Publication Data**

Robotics : design, construction and applications / Allie Weaver.
    p. cm.
Includes bibliographical references and index.
ISBN 978-1-64728-337-7
1. Robotics. 2. Robots--Design and construction. 3. Automation.
4. Machine theory. I. Weaver, Allie.
TJ211 .R63 2022
629.892--dc23

For information on all Willford Press publications
visit our website at www.willfordpress.com

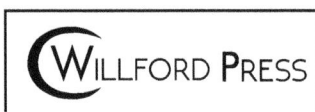

WILLFORD PRESS

# Table of Contents

| | | |
|---|---|---|
| | **Preface** | **VII** |

| | | |
|---|---|---|
| **Chapter 1** | **What is Robotics?** | **1** |
| | ▪ Developmental Robotics | 4 |
| | ▪ Beam Robotics | 7 |
| | ▪ Cloud Robotics | 11 |
| | ▪ Neurorobotics | 14 |
| | ▪ Open-source Robotics | 19 |
| | ▪ Laboratory Robotics | 20 |
| | ▪ Robotic Materials | 24 |
| | ▪ Three Laws of Robotics | 25 |
| | ▪ Aspects of Robotics | 29 |
| | ▪ Robot Kinematics | 30 |
| | ▪ Artificial Intelligence | 35 |
| | ▪ Future of Robotics | 37 |

| | | |
|---|---|---|
| **Chapter 2** | **Robots and its Types** | **39** |
| | ▪ Robots | 39 |
| | ▪ Aerobot | 49 |
| | ▪ Agricultural Robot | 53 |
| | ▪ Autonomous Robot | 57 |
| | ▪ Mobile Robot | 64 |
| | ▪ Legged Robot | 67 |
| | ▪ Hexapod Robot | 70 |
| | ▪ Humanoid Robot | 72 |
| | ▪ Medical Robot | 76 |
| | ▪ Entertainment Robot | 77 |
| | ▪ Service Robots | 79 |

| | | |
|---|---|---|
| **Chapter 3** | **Robotics: Components and Processes** | **123** |
| | ▪ Robotic Sensing | 124 |
| | ▪ Source of Power | 126 |

- Actuation                                              129
- Robotics Simulator                                     134
- Robotic Arm                                            138
- Robotic Manipulation                                   144
- Robotic Sensors                                        148

**Chapter 4  Locomotion and Control**                    **160**
- Locomotion                                             160
- Robotic Locomotion                                     160
- Bio-inspired Robotic Locomotion                        175
- Other Methods of Locomotion                            180
- Robotics Control System                                183
- Motion Planning                                        188
- Visual Servoing                                        196
- Telerobotics                                           197
- Adaptive Collaborative Control                         202
- Robot Calibration                                      206

**Chapter 5  Applications**                              **210**
- Robotic Applications in Medicine                       213
- Robot-assisted Surgery                                 216
- Robotic Applications in Agricultural Industry          223
- Human Interaction with Robots                          224
- Robot Navigation                                       227
- Different Purposes of Robots                           229
- Application of Nanorobotics                            230

**Permissions**

**Index**

# Preface

This book aims to help a broader range of students by exploring a wide variety of significant topics related to this discipline. It will help students in achieving a higher level of understanding of the subject and excel in their respective fields. This book would not have been possible without the unwavered support of my senior professors who took out the time to provide me feedback and help me with the process. I would also like to thank my family for their patience and support.

Robotics is concerned with the design, operation, construction and use of robots. It is an umbrella field which includes concepts from various disciplines such as electronic engineering, computer science, mechanical engineering and information engineering. Robotics is also involved in the usage and designing of computer systems for information processing and controlling robots. Some of the important components of robots are power course, actuators, sensors and manipulators. Robots are used in dangerous environments like bomb detection and deactivation. There are various kinds of robots such as military robots, agricultural robots, medical robots, industrial robots, collaborative robots, nanorobots, domestic robots, and autonomous drones. This book presents the complex subject of robotics in the most comprehensible and easy to understand language. Most of the topics introduced herein cover new techniques and the applications of this discipline. The book is appropriate for those seeking detailed information in this area.

A brief overview of the book contents is provided below:

Chapter – What is Robotics?

The interdisciplinary branch of engineering and science that is concerned with the design, operation, construction, and use of robots is referred to as robotics. It makes use of computer systems for their control, sensory feedback and information processing. Some of the important areas of robotics are developmental robotics, BEAM robotics, neurorobotics, laboratory robotics, etc. These diverse fields of robotics have been thoroughly discussed in this chapter.

Chapter – Robots and its Types

A robot is a machine that is capable of carrying out complex actions automatically. Some of the different types of robots are aerobot, agricultural robot, autonomous robots, mobile robots, legged robots, humanoid robots, medical robots, entertainment robots, etc. This chapter discusses these types of robots in detail.

Chapter – Robotics: Components and Processes

There are a number of components and processes used in robotics such as robotic sensing, actuation, robotics simulator, robotic arm, robotic sensors and robotic manipulation. The topics elaborated in this chapter will help in gaining a better perspective about these components and processes of robotics.

Chapter – Locomotion and Control

The collection of methods that are used by robots that enable them to transport themselves from place to place is known as robotics locomotion, legged locomotion, wheeled locomotion, bio-inspired robotic locomotion, etc. are some concepts that fall under its domain. This chapter closely examines these concepts of robotics locomotion and control to provide an extensive understanding of the subject.

Chapter – Applications

Robotics has a wide range of applications in the fields of medicine, agriculture, engineering, navigation and households. Nanorobotics is used for cancer detection and therapy, biohazard defense, etc. This chapter has been carefully written to provide an easy understanding of these applications of robotics.

**Allie Weaver**

# 1

# What is Robotics?

The interdisciplinary branch of engineering and science that is concerned with the design, operation, construction, and use of robots is referred to as robotics. It makes use of computer systems for their control, sensory feedback and information processing. Some of the important areas of robotics are developmental robotics, BEAM robotics, neurorobotics, laboratory robotics, etc. These diverse fields of robotics have been thoroughly discussed in this chapter.

Robotics is a branch of engineering and science that includes electronics engineering, mechanical engineering and computer science and so on. This branch deals with the design, construction, use to control robots, sensory feedback and information processing. These are some technologies which will replace humans and human activities in coming years. These robots are designed to be used for any purpose but these are using in sensitive environments like bomb detection, deactivation of various bombs etc. Robots can take any form but many of them have given the human appearance. The robots which have taken the form of human appearance may likely to have the walk like humans, speech, cognition and most importantly all the things a human can do. Most of the robots of today are inspired by nature and are known as bio-inspired robots.

Robotics is that branch of engineering that deals with conception, design, operation, and manufacturing of robots. There was an author named Issac Asimov, he said that he was the first person to give robotics name in a short story composed in 1940's. In that story, Issac suggested three principles about how to guide these types of robotic machines. Later on, these three principles were given the name of Issac's three laws of Robotics. These three laws state that:

- Robots will never harm human beings.

- Robots will follow instructions given by humans with breaking law one.

- Robots will protect themselves without breaking other rules.

## Creation of Robotics

The merge of numerical control and remote manipulation created a new field of engineering, and with it a number of scientific issues in design and control which are substantially different from those of the original technologies have emerged.

Robots are required to have much higher mobility and dexterity than traditional machine tools. They must be able to work in a large reachable range, access crowded places, handle a variety of workpieces, and perform flexible tasks. The high mobility and dexterity requirements result in the unique mechanical structure of robots, which parallels the human arm structure. This structure, however, significantly departs from traditional machine design. A robot mechanical structure is basically composed of cantilevered beams, forming a sequence of arm links connected by hinged joints. Such a structure has inherently poor mechanical stiffness and accuracy, hence is not appropriate for the heavy-duty, high-precision applications required of machine tools. Further, it also implies a serial sequence of servoed joints, whose errors accumulate along the linkage. In order to exploit the high mobility and dexterity uniquely featured by the serial linkage, these difficulties must be overcome by advanced design and control techniques.

The serial linkage geometry of manipulator arms is described by complex nonlinear equations. Effective analytical tools are necessary to understand the geometric and kinematic behavior of the manipulator, globally referred to as the manipulator kinematics. This represents an important and unique area of robotics research, since research in kinematics and design has traditionally focused upon single-input mechanisms with single actuators moving at constant speeds, while robots are multi-input spatial mechanisms which require more sophisticated analytical tools.

The dynamic behavior of robot manipulators is also complex, since the dynamics of multi-input spatial linkages are highly coupled and nonlinear. The motion of each joint is significantly affected by the motions of all the other joints. The inertial load imposed at each joint varies widely depending on the configuration of the manipulator arm. Coriolis and centrifugal effects are prominent when the manipulator arm moves at high speeds. The kinematic and dynamic complexities create unique control problems that are not adequately handled by standard linear control techniques, and thus make effective control system design a critical issue in robotics.

Finally, robots are required to interact much more heavily with peripheral devices than traditional numerically-controlled machine tools. Machine tools are essentially self-contained systems that handle workpieces in well-defined locations. By contrast, the environment in which robots are used is often poorly structured, and effective means must be developed to identify the locations of the workpieces as well as to communicate to peripheral devices and other machines in a coordinated fashion. Robots are also critically different from master-slave manipulators, in that they are autonomous systems. Master-slave manipulators are essentially manually controlled systems, where the human operator takes the decisions and applies control actions. The operator interprets a given task, finds an appropriate strategy to accomplish the task, and plans the procedure of operations. He/she devises an effective way of achieving the goal on the basis of his/her experience and knowledge about the task. His/her decisions are then transferred to the slave manipulator through the joystick. The resultant motion of the slave manipulator is monitored by the operator, and necessary adjustments or modifications of control actions are provided when the resultant motion is not adequate, or when unexpected events occur during the operation. The human operator is, therefore, an essential part of the control loop. When the operator is eliminated from the control system, all the planning and control commands must be generated by the machine itself. The detailed procedure of operations must be set up in advance, and each step of motion command must be generated and coded in an appropriate form so that the robot can interpret it and execute it accurately. Effective means to store the commands and manage the data file are also needed . Thus, programming and command generation are critical issues in robotics. In addition, the robot must be able to fully monitor its own motion. In order to adapt to disturbances and unpredictable changes in the work environment, the robot needs a variety of sensors, so as to obtain information both about the environment (using external sensors, such as cameras or touch sensors) and about itself (using internal sensors, such as joint encoders or joint torque sensors).

Effective sensor-based strategies that incorporate this information require advanced control algorithms. But they also imply a detailed understanding of the task.

## Manipulation and Dexterity

Contemporary industrial needs drive the applications of robots to ever more advanced tasks. Robots are required to perform highly skilled jobs with minimum human assistance or intervention. To extend the applications and abilities of robots, it becomes important to develop a sound understanding of the tasks themselves.

In order to devise appropriate arm mechanisms and to develop effective control algorithms, we need to precisely understand how a given task should be accomplished and what sort of motions the robot should be able to achieve. To perform an assembly operation, for example, we need to know how to guide the assembly part to the desired location, mate it with another part, and secure it in an appropriate way. In a grinding operation, the robot must properly position the grinding wheel while accommodating the contact force. We need to analyze the grinding process itself in order to generate appropriate force and motion commands.

A detailed understanding of the underlying principles and "know-how" involved in the task must be developed in order to use industrial robots effectively, while there is no such need for making control strategies explicit when the assembly and grinding operations are performed by a human worker. Human beings perform sophisticated manipulation tasks without being aware of the control principles involved. We have trained ourselves to be capable of skilled jobs, but in general we do not know what the acquired skills are exactly about. A sound and explicit understanding of manipulation operations, however, is essential for the long-term progress of robotics. This scientific aspect of manipulation has never been studied systematically before, and represents an emerging and important part of robotics research.

## Locomotion and Navigation

Robotics has found a number of important application areas in broad fields beyond manufacturing automation. These range from space and under-water exploration, hazardous waste disposal, and environment monitoring to robotic surgery, rehabilitation, home robotics, and entertainment. Many of these applications entail some locomotive functionality so that the robot can freely move around in an unstructured environment. Most industrial robots sit on a manufacturing floor and perform tasks in a structured environment. In contrast, those robots for non-manufacturing applications must be able to move around on their own.

Locomotion and navigation are increasingly important, as robots find challenging applications in the field. This opened up new research and development areas in robotics. Novel mechanisms are needed to allow robots to move through crowded areas, rough terrain, narrow channels, and even staircases. Various types of legged robots have been studied, since, unlike standard wheels, legs can negotiate with uneven floors and rough terrain. Among others, biped robots have been studied most extensively, resulting in the development of humanoids. Combining leg mechanisms with wheels has accomplished superior performance in both flexibility and efficiency. The Mars Rover prototype shown below has a rocker-buggy mechanism combined with advanced wheel drives in order to adapt itself to diverse terrain conditions.

Navigation is another critical functionality needed for mobile robots, in particular, for unstructured environment. Those robots are equipped with range sensors and vision system, and are capable of interpreting the data to locate themselves. Often the robot has a map of the environment, and uses it for estimating the location. Furthermore, based on the real-time data obtained in the field, the robot is capable of updating and augmenting the map, which is incomplete and uncertain in unstructured environment. As depicted, location estimation and map building are simultaneously executed in the advanced navigation system. Such Simultaneous Location and MApping (SLAM) is exactly what we human do in our daily life, and is an important functionality of intelligent robots.

The goal of robotics is thus two-fold: to extend our understanding about manipulation, locomotion, and other robotic behaviors and to develop engineering methodologies to actually perform desired tasks. The goal of this book is to provide entry-level readers and experienced engineers with fundamentals of understanding robotic tasks and intelligent behaviors as well as with enabling technologies needed for building and controlling robotic systems.

# DEVELOPMENTAL ROBOTICS

Developmental Robotics (DevRob), sometimes called epigenetic robotics, is a scientific field which aims at studying the developmental mechanisms, architectures and constraints that allow lifelong and open-ended learning of new skills and new knowledge in embodied machines. As in human children, learning is expected to be cumulative and of progressively increasing complexity, and to result from self-exploration of the world in combination with social interaction. The typical methodological approach consists in starting from theories of human and animal development elaborated in fields such as developmental psychology, neuroscience, developmental and evolutionary biology, and linguistics, then to formalize and implement them in robots, sometimes exploring extensions or variants of them. The experimentation of those models in robots allows researchers to confront them with reality, and as a consequence developmental robotics also provides feedback and novel hypotheses on theories of human and animal development.

Developmental robotics is related to, but differs from, evolutionary robotics (ER). ER uses populations of robots that evolve over time, whereas DevRob is interested in how the organization of a single robot's control system develops through experience, over time.

DevRob is also related to work done in the domains of robotics and artificial life.

## Research Directions

## Skill Domains

Due to the general approach and methodology, developmental robotics projects typically focus on having robots develop the same types of skills as human infants. A first category that is importantly being investigated is the acquisition of sensorimotor skills. These include the discovery of one's own body, including its structure and dynamics such as hand–eye coordination, locomotion, and interaction with objects as well as tool use, with a particular focus on the discovery and learning of affordances. A second category of skills targeted by developmental robots are social and linguistic skills: the acquisition of simple social behavioural games such as turn-taking, coordinated interaction, lexicons, syntax and grammar, and the grounding of these linguistic skills into sensorimotor skills (sometimes referred as symbol grounding). In parallel, the acquisition of associated cognitive skills are being investigated such as the emergence of the self/non-self distinction, the development of attentional capabilities, of categorization systems and higher-level representations of affordances or social constructs, of the emergence of values, empathy, or theories of mind.

## Mechanisms and Constraints

The sensorimotor and social spaces in which humans and robot live are so large and complex that only a small part of potentially learnable skills can actually be explored and learnt within a life-time. Thus, mechanisms and constraints are necessary to guide developmental organisms in their development and control of the growth of complexity. There are several important families of these guiding mechanisms and constraints which are studied in developmental robotics, all inspired by human development:

- Motivational systems, generating internal reward signals that drive exploration and learning, which can be of two main types:

    ○ Extrinsic motivations push robots/organisms to maintain basic specific internal properties such as food and water level, physical integrity, or light (e.g. in phototropic systems);

    ○ Intrinsic motivations push robot to search for novelty, challenge, compression or learning progress per se, thus generating what is sometimes called curiosity-driven learning and exploration, or alternatively active learning and exploration;

- Social guidance: as humans learn a lot by interacting with their peers, developmental robotics investigates mechanisms which can allow robots to participate to human-like social interaction. By perceiving and interpreting social cues, this may allow robots both to learn from humans (through diverse means such as imitation, emulation, stimulus enhancement, demonstration, etc.) and to trigger natural human pedagogy. Thus, social acceptance of developmental robots is also investigated;

- Statistical inference biases and cumulative knowledge/skill reuse: biases characterizing both representations/encodings and inference mechanisms can typically allow considerable improvement of the efficiency of learning and are thus studied. Related to this, mechanisms allowing to infer new knowledge and acquire new skills by reusing previously learnt structures is also an essential field of study;

- The properties of embodiment, including geometry, materials, or innate motor primitives/synergies often encoded as dynamical systems, can considerably simplify the acquisition of sensorimotor or social skills, and is sometimes referred as morphological computation. The interaction of these constraints with other constraints is an important axis of investigation;

- Maturational constraints: In human infants, both the body and the neural system grow progressively, rather than being full-fledged already at birth. This implies for example that new degrees of freedom, as well as increases of the volume and resolution of available sensorimotor signals, may appear as learning and development unfold. Transposing these mechanisms in developmental robots, and understanding how it may hinder or on the contrary ease the acquisition of novel complex skills is a central question in developmental robotics.

## From Bio-mimetic Development to Functional Inspiration

While most developmental robotics projects strongly interact with theories of animal and human development, the degrees of similarities and inspiration between identified biological mechanisms and their counterpart in robots, as well as the abstraction levels of modeling, may vary a lot. While some projects aim at modeling precisely both the function and biological implementation (neural or morphological models), such as in neurorobotics, some other projects only focus on functional modeling of the mechanisms and constraints described above, and might for example reuse in their architectures techniques coming from applied mathematics or engineering fields.

## Objectives

As developmental robotics is a relatively novel research field and at the same time very ambitious, many fundamental open challenges remain to be solved.

First of all, existing techniques are far from allowing real-world high-dimensional robots to learn an open- ended repertoire of increasingly complex skills over a life-time period. High-dimensional continuous sensorimotor spaces are a major obstacle to be solved. Lifelong cumulative learning is another one. Actually, no experiments lasting more than a few days have been set up so far, which contrasts severely with the time period needed by human infants to learn basic sensorimotor skills while equipped with brains and morphologies which are tremendously more powerful than existing computational mechanisms.

Among the strategies to explore in order to progress towards this target, the interaction between the mechanisms and constraints described in the previous section shall be investigated more systematically. Indeed, they have so far mainly been studied in isolation. For example, the interaction of intrinsically motivated learning and socially guided learning, possibly constrained by maturation, is an essential issue to be investigated.

Another important challenge is to allow robots to perceive, interpret and leverage the diversity of multimodal social cues provided by non-engineer humans during human-robot interaction. These capacities are so far mostly too limited to allow efficient general purpose teaching from humans.

A fundamental scientific issue to be understood and resolved, which applied equally to human development, is how compositionality, functional hierarchies, primitives, and modularity, at all levels of sensorimotor and social structures, can be formed and leveraged during development. This is deeply linked with the problem of the emergence of symbols, sometimes referred as the "symbol grounding problem" when it comes to language acquisition. Actually, the very existence and need for symbols in the brain is actively questioned, and alternative concepts, still allowing for compositionality and functional hierarchies are being investigated.

During biological epigenesis, morphology is not fixed but rather develops in constant interaction with the development of sensorimotor and social skills. The development of morphology poses obvious practical problems with robots, but it may be a crucial mechanism that should be further explored, at least in simulation, such as in morphogenetic robotics.

Another open problem is the understanding of the relation between the key phenomena investigated by developmental robotics (e.g., hierarchical and modular sensorimotor systems, intrinsic/extrinsic/social motivations, and open-ended learning) and the underlying brain mechanisms.

Similarly, in biology, developmental mechanisms (operating at the ontogenetic time scale) strongly interact with evolutionary mechanisms (operating at the phylogenetic time scale) as shown in the flourishing "evo-devo" scientific literature. However, the interaction of those mechanisms in artificial organisms, developmental robots in particular, is still vastly understudied. The interaction of evolutionary mechanisms, unfolding morphologies and developing sensorimotor and social skills will thus be a highly stimulating topic for the future of developmental robotics.

# BEAM ROBOTICS

BEAM robotics is a style of robotics that primarily uses simple analogue circuits, such as comparators, instead of a microprocessor in order to produce an unusually simple design. While not as flexible as microprocessor based robotics, BEAM robotics can be robust and efficient in performing the task for which it was designed.

BEAM robots may use a set of the analog circuits, mimicking biological neurons, to facilitate the robot's response to its working environment.

## Mechanisms and Principles

The basic BEAM principles focus on a stimulus-response based ability within a machine. The underlying mechanism was invented by Mark W. Tilden where the circuit (or a Nv net of Nv neurons) is used to simulate biological neuron behaviours. Some similar research was previously done by Ed Rietman in ‹Experiments In Artificial Neural Networks'. Tilden's circuit is often compared to a shift register, but with several important features making it a useful circuit in a mobile robot.

Other rules that are included (and to varying degrees applied):

1. Use the lowest number possible of electronic elements (*"keep it simple"*).
2. Recycle and reuse technoscrap.
3. Use radiant energy (such as solar power).

There are a large number of BEAM robots designed to use solar power from small solar arrays to power a "Solar Engine" which creates autonomous robots capable of operating under a wide range of lighting conditions. Besides the simple computational layer of Tilden's "Nervous Networks", BEAM has brought a multitude of useful tools to the roboticist's toolbox. The "Solar Engine" circuit, many H-bridgecircuits for small motor control, tactile sensor designs, and meso-scale (palm-sized) robot construction techniques have been documented and shared by the BEAM community.

## BEAM Robots

Being focused on "reaction-based" behaviors (as originally inspired by the work of Rodney Brooks), BEAM robotics attempts to copy the characteristics and behaviours of biological organisms, with the ultimate goal of domesticating these "wild" robots. The aesthetics of BEAM robots derive from the principle "form follows function" modulated by the particular design choices the builder makes while implementing the desired functionality.

## Disputes in the Name

Various people have varying ideas about what BEAM actually stands for. The most widely accepted meaning is Biology, Electronics, Aesthetics, and Mechanics.

This term originated with Mark Tilden during a discussion at the Ontario Science Centre in 1990. Mark was displaying a selection of his original bots which he had built while working at the University of Waterloo.

However, there are many other semi-popular names in use:

- Biotechnology Ethology Analogy Morphology.

- Building Evolution Anarchy Modularity.

### Microcontrollers

Unlike many other types of robots controlled by microcontrollers, BEAM robots are built on the principle of using multiple simple behaviours linked directly to sensor systems with little signal conditioning. This design philosophy is closely echoed in the classic book "Vehicles: Experiments in Synthetic Psychology". Through a series of thought experiments, this book explores the development of complex robot behaviours through simple inhibitory and excitory sensor links to the actuators. Microcontrollers and computer programming are usually not a part of a traditional (aka., "pure" ) BEAM robot due to the very low-level hardware-centric design philosophy.

There are successful robot designs mating the two technologies. These "hybrids" fulfill a need for robust control systems with the added flexibility of dynamic programming, like the "horse-and-rider" topology BEAMbots (e.g. the ScoutWalker 3 ). 'Horse' behavior is implemented with traditional BEAM technology but a microcontroller based 'rider' can guide that behavior so as to accomplish the goals of the 'rider'.

## Types

There are various "-*trope*" BEAMbots, which attempt to achieve a specific goal. Of the series, the phototropes are the most prevalent, as light-seeking would be the most beneficial behaviour for a solar-powered robot.

- Audiotropes react to sound sources.

  ◦ *Audiophiles* go towards sound sources.

  ◦ *Audiophobes* go away from sound sources.

- Phototropes ( "light-seekers") react to light sources.
  - *Photophiles* (also *Photovores*) go toward light sources.
  - *Photophobes* go away from light sources.
- Radiotropes react to radio frequency sources.
  - *Radiophiles* go toward RF sources.
  - *Radiophobes* go away from RF sources.
- Thermotropes react to heat sources.
  - *Thermophiles* go toward heat sources.
  - *Thermophobes* go away from heat sources.

BEAMbots have a variety of movements and positioning mechanisms. These include:

- *Sitters*: Unmoving robots that have a physically passive purpose.
  - Beacons: Transmit a signal (usually a navigational blip) for other BEAMbots to use.
  - Pummers: Display a "light show".
  - Ornaments: A catch-all name for sitters that are not beacons or pummers.
- *Squirmers*: Stationary robots that perform an interesting action (usually by moving some sort of limbs or appendages).
  - Magbots: Use magnetic fields for their mode of animation.
  - Flagwavers: Move a display (or "flag") around at a certain frequency.
  - Heads: Pivot and follow some detectable phenomena, such as a light (These are popular in the BEAM community. They can be stand-alone robots, but are more often incorporated into a larger robot.).
  - Vibrators: Use a small pager motor with an off-centre weight to shake themselves about.
- *Sliders*: Robots that move by sliding body parts smoothly along a surface while remaining in contact with it.
  - Snakes: Move using a horizontal wave motion.
  - Earthworms: Move using a longitudinal wave motion.
- *Crawlers*: Robots that move using tracks or by rolling the robot's body with some sort of appendage. The body of the robot is not dragged on the ground.
  - Turbots: Roll their entire bodies using their arm(s) or flagella.
  - Inchworms: Move part of their bodies ahead, while the rest of the chassis is on the ground.

- Tracked robots: Use tracked wheels, like a tank.
- *Jumpers*: Robots which propel themselves off the ground as a means of locomotion.
  - Vibrobots: Produce an irregular shaking motion moving themselves around a surface.
  - Springbots: Move forward by bouncing in one particular direction.
- *Rollers*: Robots that move by rolling all or part of their body.
  - Symets: Driven using a single motor with its shaft touching the ground, and moves in different directions depending on which of several symmetric contact points around the shaft are touching the ground.
  - Solarrollers: Solar-powered cars that use a single motor driving one or more wheels; often designed to complete a fairly short, straight and level course in the shortest amount of time.
  - Poppers: Use two motors with separate solar engines; rely on differential sensors to achieve a goal.
  - Miniballs: Shift their centre of mass, causing their spherical bodies to roll.
- *Walkers*: Robots that move using legs with differential ground contact.
  - Motor Driven: Use motors to move their legs (typically 3 motors or less).
  - Muscle Wire Driven: use Nitinol (nickel - titanium alloy) wires for their leg actuators.
- *Swimmers*: Robots that move on or below the surface of a liquid (typically water).
  - Boatbots: Operate on the surface of a liquid.
  - Subbots: Operate under the surface of a liquid.
- *Fliers*: Robots that move through the air for sustained periods.
  - Helicopters: Use a powered rotor to provide both lift and propulsion.
  - Planes: Use fixed or flapping wings to generate lift.
  - Blimps: Use a neutrally-buoyant balloon for lift.
- *Climbers*: Robot that moves up or down a vertical surface, usually on a track such as a rope or wire.

## Applications and Current Progress

At present, autonomous robots have seen limited commercial application, with some exceptions such as the iRobot Roomba robotic vacuum cleaner and a few lawn-mowing robots. The main practical application of BEAM has been in the rapid prototyping of motion systems and hobby/education applications. Mark Tilden has successfully used BEAM for the prototyping of products for Wow-Wee Robotics, as evidenced by B.I.O.Bug and RoboRaptor. Solarbotics Ltd., Bug'n'Bots, JCM InVentures Inc., and PagerMotors.com have also brought BEAM-related hobby and educational goods to the marketplace. Vex has also developed Hexbugs, tiny BEAM robots.

Aspiring BEAM roboticists often have problems with the lack of direct control over "pure" BEAM control circuits. There is ongoing work to evaluate biomorphic techniques that copy natural systems because they seem to have an incredible performance advantage over traditional techniques. There are many examples of how tiny insect brains are capable of far better performance than the most advanced microelectronics.

Another barrier to widespread application of BEAM technology is the perceived random nature of the 'nervous network', which requires new techniques to be learned by the builder to successfully diagnose and manipulate the characteristics of the circuitry. A think-tank of international academics meet annually in Telluride, Colorado to address this issue directly, and until recently, Mark Tilden has been part of this effort (he had to withdraw due to his new commercial commitments with Wow-Wee toys).

Having no long-term memory, BEAM robots generally do not learn from past behaviour. However, there has been work in the BEAM community to address this issue. One of the most advanced BEAM robots in this vein is Bruce Robinson's Hider, which has an impressive degree of capability for a microprocessor-less design.

# CLOUD ROBOTICS

Cloud robotics is a field of robotics that attempts to invoke cloud technologies such as cloud computing, cloud storage, and other Internet technologies centred on the benefits of converged infrastructure and shared services for robotics. When connected to the cloud, robots can benefit from the powerful computation, storage, and communication resources of modern data center in the cloud, which can process and share information from various robots or agent (other machines, smart objects, humans, etc.). Humans can also delegate tasks to robots remotely through networks. Cloud computing technologies enable robot systems to be endowed with powerful capability whilst reducing costs through cloud technologies. Thus, it is possible to build lightweight, low cost, smarter robots have intelligent "brain" in the cloud. The "brain" consists of data center, knowledge base, task planners, deep learning, information processing, environment models, communication support, etc.

## Components

A cloud for robots potentially has at least six significant components:

- Offering a global library of images, maps, and object data, often with geometry and mechanical properties, expert system, knowledge base (i.e. semantic web, data centres);

- Massively-parallel computation on demand for sample-based statistical modelling and motion planning, task planning, multi-robot collaboration, scheduling and coordination of system;

- Robot sharing of outcomes, trajectories, and dynamic control policies and robot learning support;

- Human sharing of "open-source" code, data, and designs for programming, experimentation, and hardware construction;

- On-demand human guidance and assistance for evaluation, learning, and error recovery;

- Augmented human–robot interaction through various way (Semantics knowledge base, Apple SIRI like service etc.).

## Applications

### Autonomous Mobile Robots

Google's self-driving cars are cloud robots. The cars use the network to access Google's enormous database of maps and satellite and environment model (like Streetview) and combines it with streaming data from GPS, cameras, and 3D sensors to monitor its own position within centimetres, and with past and current traffic patterns to avoid collisions. Each car can learn something about environments, roads, or driving, or conditions, and it sends the information to the Google cloud, where it can be used to improve the performance of other cars.

### Cloud Medical Robots

a medical cloud (also called a healthcare cluster) consists of various services such as a disease archive, electronic medical records, a patient health management system, practice services, analytics services, clinic solutions, expert systems, etc. A robot can connect to the cloud to provide clinical service to patients, as well as deliver assistance to doctors (e.g. a co-surgery robot). Moreover, it also provides a collaboration service by sharing information between doctors and care givers about clinical treatment.

### Assistive Robots

A domestic robot can be employed for healthcare and life monitoring for elderly people. The system collects the health status of users and exchange information with cloud expert system or doctors to facilitate elderly peoples life, especially for those with chronic diseases. For example, the robots are able to provide support to prevent the elderly from falling down, emergency healthy support such as heart disease, blooding disease. Care givers of elderly people can also get notification when in emergency from the robot through network.

### Industrial Robots

As highlighted by the German government's Industry 4.0 Plan, "Industry is on the threshold of the fourth industrial revolution. Driven by the Internet, the real and virtual worlds are growing closer and closer together to form the Internet of Things. Industrial production of the future will be characterised by the strong individualisation of products under the conditions of highly flexible (large series) production, the extensive integration of customers and business partners in business and value-added processes, and the linking of production and high-quality services leading to so-called hybrid products." In manufacturing, such cloud based robot systems could learn to handle tasks such as threading wires or cables, or aligning gaskets from a professional knowledge base. A group of robots can share information for some collaborative tasks. Even more, a consumer is able to place customised product orders to manufacturing robots directly with online ordering systems. Another potential paradigm is shopping-delivery robot systems. Once an order is placed, a warehouse robot dispatches the item to an autonomous car or autonomous drone to delivery it to its recipient.

## Limitations of Cloud Robotics

Though robots can benefit from various advantages of cloud computing, cloud is not the solution to all of robotics.

- Controlling a robot's motion which relies heavily on (real-time) sensors and feedback of controller may not benefit much from the cloud.

- Tasks that involve real-time execution require on-board processing.

- Cloud-based applications can get slow or unavailable due to high-latency responses or network hitch. If a robot relies too much on the cloud, a fault in the network could leave it "brainless".

## Challenges

The research and development of cloud robotics has following potential issues and challenges:

- Scalable parallelisation-grid-computing, parallelisation schemes scale with the size of automation infrastructure.

- Effective load balancing: Balancing operations between local and cloud computation.

- Knowledge bases and representations.

- Collective learning for automation in cloud.

- Infrastructure/Platform or Software as a Service.

- Internet of Things for robotics.

- Integrated and collaborative fault-tolerant control.

- Big Data: Data, collected and/or disseminated over large, accessible networks can enable decisions for classification problems or reveal patterns.

- Wireless communication, Connectivity to the cloud.

- System architectures of robot cloud.

- Open-source, open-access infrastructures.

- Workload-sharing.

- Standards and Protocols.

## Risks

- Environmental security - The concentration of computing resources and users in a cloud computing environment also represents a concentration of security threats. Because of their size and significance, cloud environments are often targeted by virtual machines and bot malware, brute force attacks, and other attacks.

- Data privacy and security - Hosting confidential data with cloud service providers involves the transfer of a considerable amount of an organisation's control over data security to the provider. For example, every cloud contains a huge information from the clients include personal data. If a household robot is hacked, users could have risk of their personal privacy and security, like house layout, life snapshot, home-view, etc. It may be accessed and leaked to the world around by criminals. Another problems is once a robot is hacked and controlled by someone else, which may put the user in danger.

- Ethical problems - Some ethics of robotics, especially for cloud based robotics must be considered. Since a robot is connected via networks, it has risk to be accessed by other people. If a robot is out of control and carries out illegal activities, who should be responsible for it.

# NEUROROBOTICS

Neurorobots are robotic devices that have control systems based on principles of the nervous system. These models operate on the premise that the "brain is embodied and the body is embedded in the environment". Therefore, neurorobots are grounded and situated in a real environment. The real environment is required for two reasons. First, simulating an environment can introduce unwanted and unintentional biases to the model. For example, a computer-generated object presented to a vision model has its shape and segmentation defined by the modeler and directly presented to the model, whereas a device that views an object hanging on a wall has to discern the shape and figure from ground segmentation based on its on active vision. Second, real environments are rich, multimodal, and noisy; an artificial design of such an environment would be computationally intensive and difficult to simulate. However, all these interesting features of the environment come for "free" when a neurorobot is placed in the real world. The field of Neurorobotics started in the late 1980s. Kawato and colleagues built a series of robotic devices to test how the cerebellum adapts movements. Gerald Edelman's group tested the Theory of Neuronal Group Selection by introducing the Darwin series of automata. Since this time, the number of neuroroboticists has expanded into a full community of researchers studying a wide-range of neuroscience topics.

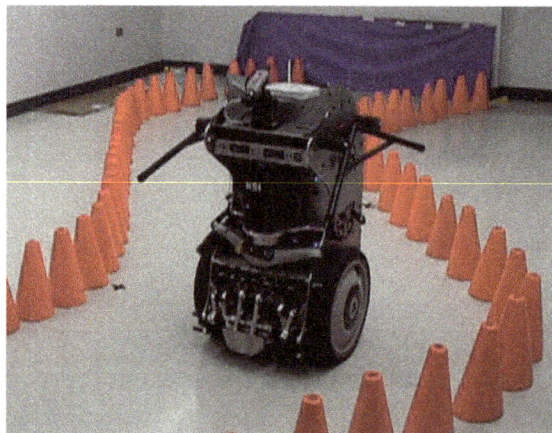

A brain-based device with a simulated cerebellum for predictive motor control. The device is built on the Segway Robotic Mobility Platform. The device navigated a path dictated by the orange traffic cones that were spaced a few inches apart. The BBD's task was to traverse a curved course outlined by traffic cones, without collisions. Initially, collisions or near collisions with the cones generated a reflexive movement away from the obstacle and a reflexive braking response. These reflex commands were also used as error signals to the cerebellar model via simulated climbing fiber inputs. Success in this task required the BBD's cerebellum to associate predictive visual motion cues, which came from optic flow generated by self-movement, with the correct movements to avoid collisions with the cone boundaries.

A neurorobot has the Following Properties:

1. It engages in a behavioral task.

2. It is situated in a real-world environment.

3. It has a means to sense environmental cues and act upon its environment.

4. Its behavior is controlled by a simulated nervous system having a design that reflects, at some level, the brain's architecture and dynamics.

As a result of these properties, neurorobotic models provide heuristics for developing and testing theories of brain function in the context of phenotypic and environmental interactions. Also, neurorobotic models may provide a foundation for the development of more effective robots, based on an improved understanding of the biological bases of adaptive behavior.

## Classes of Neurorobotic Models

There are too many examples of neurobiologically inspired robotic devices. However, the approach has been applied to several distinct areas of neuroscience research:

1. Motor control and locomotion.

2. Learning and memory systems.

3. Value systems and action selection.

## Motor Control and Locomotion

Neurorobots have proved useful for investigating animal locomotion and motor control, and for designing robot controllers. Neural models of central pattern generators, pools of motorneurons that drive a repetitive behavior, have been used to control locomotion in robots. Kimura and colleagues have shown how neurorobotics can provide a bridge between neuroscience and biomechanics by demonstrating emergent 4-legged locomotion based on central pattern generator mechanisms modulated by reflexes. Their group developed a model of a learnable pattern generator and demonstrated its viability using a series of synthetic and humanoid robotic examples. Ijspeert and colleagues constructed an amphibious salamander-like robot that is capable of both swimming and walking, and therefore represents a key stage in the evolution of vertebrate-legged locomotion. A neurorobotic implementation was found necessary for (1) testing whether the models could produce locomotion both in water and on ground and (2) investigating how sensory feedback affects dynamic pattern generation.

An intriguing neural inspiration for the design of robot controllers is the mirror neuron system found in primates. Mirror neurons in the premotor cortex are active, both when a monkey grasps or manipulates objects and when it watches another animal performing similar actions. Neuro-roboticists, using this notion of mirror neurons, have suggested that complex movements such as reaching and locomotion may be achieved through imitation.

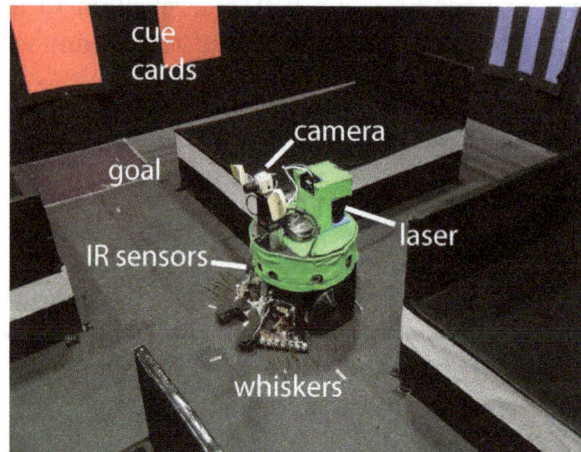

Figure: Darwin XI, a brain-based device with a simulated hippocampus and its surrounding regions. Darwin XI is pictured at the choice point of its plus-maze environment. Darwin XI began a trial alternately at the East or West start arm, and used its artificial whiskers to follow the maze arm until it reached the choice point. As it followed the maze wall its whiskers sensed patterns of pegs, its camera sensed color cue cards on the perimeter, its compass provided heading, and its laser provided range information. In the beginning of training, Darwin XI was given a rewarding stimulus when it chose the South goal arm. After it successfully learned that task, the rewarding stimulus was switched to the North goal arm.

Another strategy for motor control in neurally inspired robots is to use a predictive controller to convert awkward, error prone movements into smooth, accurate movements. Recent theories of motor control suggest that the cerebellum learns to replace primitive reflexes with predictive motor signals. The idea is that the outcomes of reflexive motor commands provide error signals for a predictive controller, which then learns to produce a correct motor control signal prior to the less adaptive reflex response. Neurally inspired models have used these ideas in the design of robots that learn to avoid obstacles, produce accurate eye and generate adaptive arm movements. Figure shows a brain-based device, containing a model of the cerebellum and cortical area MT, which learned to predict collisions based on visual motion cues and adapted its movements accordingly.

## Learning and Memory Systems

A major theme in neurorobotics is neurally inspired models of learning and memory. One area of particular interest is navigation systems based on the rodent hippocampus. Rats have exquisite navigation capabilities in both the light and in the dark. Moreover, the finding of place cells in the rodent hippocampus, which fire specifically at a spatial location, have been of theoretical interest for models of memory and route planning. Robots with models of the hippocampal place cells have been shown to be viable for navigation in mazes and environments similar to those used in rat spatial memory studies. Recently, large-scale systems-level models of the hippocampus and its surrounding regions have been embedded on robots to investigate the role of these regions in the acquisition and recall of episodic memory . Figure shows a brain-based device in a plus maze that developed episodic-like responses in its simulated hippocampus.

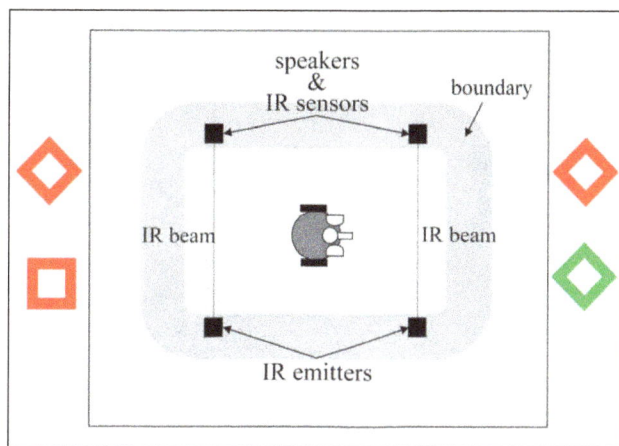

Darwin VIII views objects on two of the walls of an arena. When Darwin VIII breaks the beam from an IR emitter to an IR sensor, a tone is emitted from a speaker on the side of the red diamond. The tone triggers Darwin VIII's value system and causes it to associate value with the red diamond.

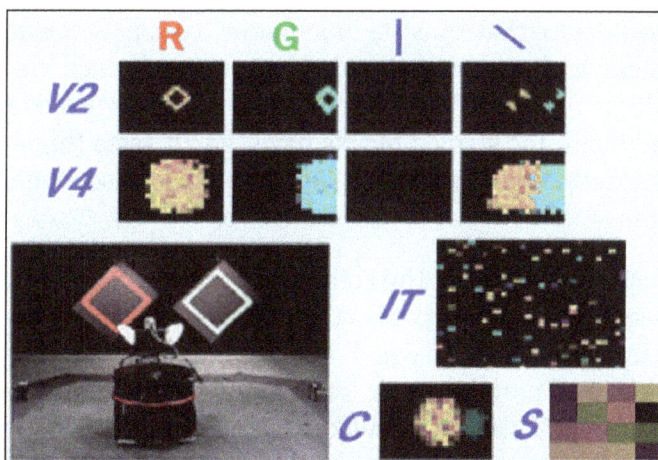

Snapshot of Darwin VIII's neuronal unit activity during a behavioral experiment. Each pixel in the neural areas represents a neuronal unit; the activity is normalized from no activity (black) to maximum activity (bright colors), and the phase (i.e. timing of activity) is indicated by the color of the pixel. The neuronal units responding to the attributes of the red diamond share a common phase (red-orange color), whereas the neuronal units responding to the green diamond share a different phase (blue-green color).

Another learning and memory property that is of importance to the development of neurorobotics is the ability to organize the unlabeled signals that robots receive from the environment into categories. This organization of signals, which in general depends on a combination of sensory modalities (e.g. vision, sound, taste, or touch), is called perceptual categorization. Several neurorobots have been constructed that build up such categories, without instruction, by combining auditory, tactile, taste, and visual cues from the environment. Figure show a brain-based device that developed categories for the objects it observed and solved the visual binding problem through synchronous activity in its simulated ventral visual stream. These categories emerged from the device's experience exploring its environment.

## Value Systems and Action Selection

Darwin VII, a brain-based device that consists of a mobile base, a CCD camera, two microphones on either side of the camera, and sensors embedded in a gripper, which measures the surface conductivity of the metal blocks it manipulates. These sensory signals provide input to the neuronal simulation. In this experiment, the striped blocks have "good" taste (highly conductive), and the spotted blocks have "bad" taste (weakly conductive). The blocks also would emit tones for an auditory cue. Adapted from.

Biological organisms adapt their behavior through value systems, which provide nonspecific, modulatory signals to the rest of the brain that bias the outcome of local changes in synaptic efficacy in the direction needed to satisfy global needs. Examples of value systems in the brain include the dopaminergic, cholinergic, and noradrenergic systems. Behavior that evokes positive responses in value systems biases synaptic change to make production of the same behavior more likely when the situation in the environment (and thus the local synaptic inputs) is similar; behavior that evokes negative value biases synaptic change in the opposite direction. The dopamine system and its role in shaping decision making has been explored in neurorobots and brain-based devices. Figure shows a brain-based device that learned to associate a neutral stimulus (i.e. visual category) with an innate value (i.e. conductivity of metal blocks). Doya's group has been investigating the effect of multiple neuromodulators in the "Cyber-rodent"; two-wheeled robots that move autonomously in an environment. These robots have drives for self-preservation and self-reproduction exemplified by searching for and recharging from battery packs on the floor and then communicating this information to other robots nearby through their infrared communication ports. In addition to examining how neuromodulators such as dopamine can influence decision making, neuroroboticists have been investigating the basal ganglia as a model that mediates action selection. Based on the architecture of the basal ganglia, Prescott and colleagues embedded a model of the basal ganglia in a robot that had to select from several actions depending on the environmental context.

## Conclusions

Higher brain functions depend on the cooperative activity of an entire nervous system, reflecting its morphology, its dynamics, and its interaction with the environment. Neurorobots are designed to incorporate these attributes such that they can test theories of brain function. The behavior of neurorobots and the activity of their simulated nervous systems allow for comparisons with experimental data acquired from animals. The comparison can be made at the behavioral level, the systems level, and the neuronal level. These comparisons serve two purposes: First, neurorobots can generate hypotheses and test theories of brain function. The construction of a complete behaving model forces the designer to specify theoretical and implementation details that can be easy to overlook in an ungrounded or disembodied theoretical model. Moreover, it forces these details to be consistent. Second, by using the animal nervous system as a metric, neurorobot designers can continually make their simulated nervous systems and resulting behavior closer to that of the model animal. This, in turn, allows the eventual creation of practical devices that may approach the sophistication of living organisms.

# OPEN-SOURCE ROBOTICS

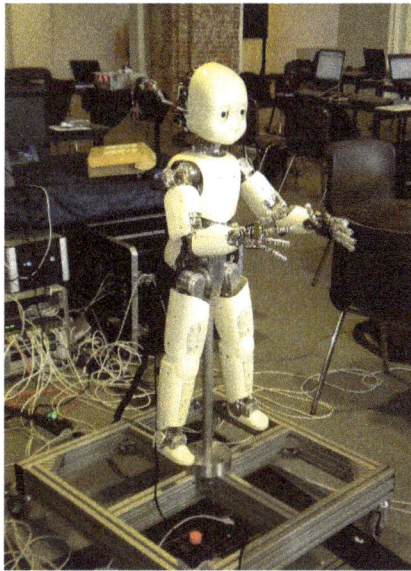

An open source iCub robot mounted on a supporting frame. The robot is 104 cm high and weighs around 22 kg.

Open-source robotics (OSR) is where the physical artifacts of the subject are offered by the open design movement. This branch of robotics makes use of open-source hardware and free and open-source software providing blueprints, schematics, and source code. The term usually means that information about the hardware is easily discerned so that others can make it from standard commodity components and tools—coupling it closely to the maker movement and open science.

## Advantages

- Long-term availability: Many non-open robots and components, especially at hobbyist level, are designed and sold by tiny startups which can disappear overnight, leaving customers without support. Open-source systems are guaranteed to have their designs available for ever so communities of users can, and do, continue support after the manufacturer has disappeared.

- Avoiding lock-in: A company relying on any particular non-open component exposes itself to business risk that the supplier could ratchet up prices after they have invested time and technology building on it. Open hardware can be manufacturered by anyone, creating competition or at least the potential for competition, which both remove this risk.

- Interchangeable software and/or hardware with common interfaces.

- Ability to modify and fork designs more easily for customisation.

- Scientific reproducibility guarantees that other labs can replicate and extend work, leading to increased impact, citations and reputation for the designer.

- Lower-cost: Costs of a robot can be decreased dramatically when all components and tools are commodities. No component seller can hold a project to ransom by ratcheting the price of a critical component, as competing suppliers can easily be interchanged.

## Drawbacks

- For commercial organisations, open-sourcing their own designs obviously means they can no longer make large profits through the traditional engineering business model of acting as the monopoly manufacturer or seller, because the open design can be manufactured and sold by anyone including direct competitors. Profit from engineering can come from three main sources: design, manufacturing, and support. As with other open source business models, commercial designers typically make profit via their association with the brand, which may still be trademarked. A valuable brand allows them to command a premium for their own manufactured products, as it can be associated with high quality and provide a quality guarantee to customers. The same brand is also used to command a premium on associated services, such as providing installation, maintenance, and integration support for the product. Again customers will typically pay more for the knowledge that this support is provided directly by the original designer, who therefore knows the product better than competitors.

- Some customers associate open source with amateurism, the hacker community, low quality and poor support. Serious companies using this business model may need to work harder to overcome this perception by emphasising their professionalism and brand to differentiate themselves from amateur efforts.

# LABORATORY ROBOTICS

Laboratory robotics is the act of using robots in biology or chemistrylabs. For example, pharmaceutical companies employ robots to move biological or chemical samples around to synthesize novel chemical entities or to test pharmaceutical value of existing chemical matter. Advanced laboratory robotics can be used to completely automate the process of science, as in the Robot Scientist project.

Laboratory robots doing acid digestion chemical analysis.

Laboratory processes are suited for robotic automation as the processes are composed of repetitive movements (e.g. pick/place, liquid & solid additions, heating/cooling, mixing, shaking, testing). Many laboratory robots are commonly referred as autosamplers, as their main task is to provide continuous samples for analytical devices.

## Applications

### Low-cost Laboratory Robotics

Low-cost robotic arm used as an autosampler.

The high cost of many laboratory robots has inhibited their adoption. However, currently there are many robotic devices that have very low cost, and these could be employed to do some jobs in a laboratory. For example, a low-cost robotic arm was employed to perform several different kinds of water analysis, without loss of performance compared to much more expensive autosamplers. Alternatively, the autosampler of a device can be used with another device, thus avoiding the need for purchasing a different autosampler or hiring a technician for doing the job. The key aspects to achieve low-cost in laboratory robotics are 1) the use of low-cost robots, which become more and more common, and 2) the use of scripting, which enables compatibility between robots and other analytical equipment.

## Biological Laboratory Robotics

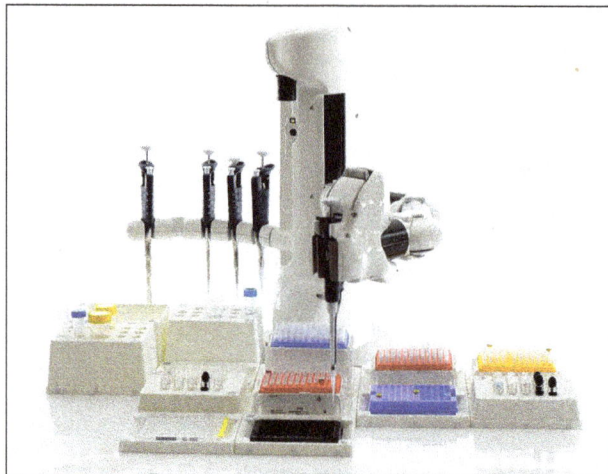

An example of pipettes and microplates manipulated by an anthropomorphic robot.

Biological and chemical samples, in either liquid or solid state, are stored in vials, plates or tubes. Often, they need to be frozen and/or sealed to avoid contamination or to retain their biological and/or chemical properties. Specifically, the life science industry has standardized on a plate format, known as the microtiter plate, to store such samples.

The microtiter plate standard was formalized by the Society for Biomolecular Screening in 1996. It typically has 96, 384 or even 1536 sample wells arranged in a 2:3 rectangular matrix. The standard governs well dimensions (e.g. diameter, spacing and depth) as well as plate properties (e.g. dimensions and rigidity).

A number of companies have developed robots to specifically handle SBS microplates. Such robots may be liquid handlers which aspirates or dispenses liquid samples from and to these plates, or "plate movers" which transport them between instruments.

Other companies have pushed integration even further: on top of interfacing to the specific consumables used in biology, some robots have been designed with the capability of interfacing to volumetric pipettes used by biologists and technical staff. Essentially, all the manual activity of liquid handling can be performed automatically, allowing humans spending their time in more conceptual activities.

Instrument companies have designed plate readers which can carry out detect specific biological, chemical or physical events in samples stored in these plates. These readers typically use optical and/or computer vision techniques to evaluate the contents of the microtiter plate wells.

One of the first applications of robotics in biology was peptide and oligonucleotide synthesis. One early example is the polymerase chain reaction (PCR) which is able to amplify DNA strands using a thermal cycler to micromanage DNA synthesis by adjusting temperature using a pre-made computer program. Since then, automated synthesis has been applied to organic chemistry and expanded into three categories: reaction-block systems, robot-arm systems, and non-robotic fluidic systems. The primary objective of any automated workbench is high-throughput processes and cost reduction. This allows a synthetic laboratory to operate with a fewer number of people working more efficiently.

## Pharmaceutical Applications

One major area where automated synthesis has been applied is structure determination in pharmaceutical research. Processes such as NMR and HPLC-MS can now have sample preparation done by robotic arm. Additionally, structural protein analysis can be done automatically using a combination of NMR and X-ray crystallography. Crystallization often takes hundreds to thousands of experiments to create a protein crystal suitable for X-ray crystallography. An automated micropipet machine can allow nearly a million different crystals to be created at once, and analyzed via X-ray crystallography.

## Combinatorial Library Synthesis

Robotics have applications with Combinatorial Chemistry which has great impact on the pharmaceutical industry. The use of robotics has allowed for the use of much smaller reagent quantities and mass expansion of chemical libraries. The "parallel synthesis" method can be improved upon with automation. The main disadvantage to "parallel-synthesis" is the amount of time it takes to develop a library, automation is typically applied to make this process more efficient.

The main types of automation are classified by the type of solid-phase substrates, the methods for adding and removing reagents, and design of reaction chambers. Polymer resins may be used as a substrate for solid-phase. It is not a true combinatorial method in the sense that "split-mix" where a peptide compound is split into different groups and reacted with different compounds. This is then mixed back together split into more groups and each groups is reacted with a different compound. Instead the "parallel-synthesis" method does not mix, but reacts different groups of the same peptide with different compounds and allows for the identification of the individual compound on each solid support. A popular method implemented is the reaction block system due to its relative low cost and higher output of new compounds compared to other "parallel-synthesis" methods. Parallel-Synthesis was developed by Mario Geysen and his colleagues and is not a true type of combinatorial synthesis, but can be incorporated into a combinatorial synthesis. This group synthesized 96 peptides on plastic pins coated with a solid support for the solid phase peptide synthesis. This method uses a rectangular block moved by a robot so that reagents can be pipetted by a robotic pipetting system. This block is separated into wells which the individual reactions take place. These compounds are later cleaved from the solid-phase of the well for further analysis. Another method is the closed reactor system which uses a completely closed off reaction vessel with a series of fixed connections to dispense. Though the produce fewer number of compounds than other methods, its main advantage is the control over the reagents and reaction conditions. Early closed reaction systems were developed for peptide synthesis which required variations in temperature and a diverse range of reagents. Some closed reactor system robots have a temperature range of 200 °C and over 150 reagents.

## Purification

Simulated distillation, a type of gas chromatography testing method used in the petroleum, can be automated via robotics. An older method used a system called ORCA (Optimized Robot for Chemical Analysis)was used for the analysis of petroleum samples by simulated distillation (SIMDIS). ORCA has allowed for shorter analysis times and has reduced maximum temperature needed to elute compounds. One major advantage of automating purification is the scale at which separations can be done. Using microprocessors, ion-exchange separation can be conducted on a nanoliter scale in a short period of time.

Robotics have been implemented in liquid-liquid extraction (LLE) to streamline the process of preparing biological samples using 96-well plates. This is an alternative method to solid-phase extraction methods and protein precipitation, which has the advantage of being more reproducible and robotic assistance has made LLE comparable in speed to solid phase extraction. The robotics used for LLE can perform an entire extraction with quantities in the microliter scale and performing the extraction in as little as ten minutes.

## Advantages and Disadvantages

### Advantages

One of the advantages to automation is faster processing, but it is not necessarily faster than a human operator. Repeatability and reproducibility are improved as automated systems as less likely to have variances in reagent quantities and less likely to have variances in reaction conditions. Typically productivity is increased since human constraints, such as time constraints, are no longer a factor. Efficiency is generally improved as robots can work continuously and reduce the amount of reagents used to perform a reaction. Also there is a reduction in material waste. Automation can also establish safer working environments since hazardous compounds do not have to be handled. Additionally automation allows staff to focus on other tasks that are not repetitive.

### Disadvantages

Typically the cost of a single synthesis or sample assessment are expensive to set up and start up cost for automation can be expensive. Many techniques have not been developed for automation yet. Additionally there is difficultly automating instances where visual analysis, recognition, or comparison is required such as color changes. This also leads to the analysis being limited by available sensory inputs. One potential disadvantage is an increases job shortages as automation may replace staff members who do tasks easily replicated by a robot. Some systems require the use of programming languages such as C++ or Visual Basicto run more complicated tasks.

# ROBOTIC MATERIALS

Robotic materials are composite materials that combine sensing, actuation, computation, and communication in a repeatable or amorphous pattern. Robotic materials can be considered computational metamaterials in that they extend the original definition of a metamaterial as "macroscopic composites having a man-made, three-dimensional, periodic cellular architecture designed to produce an optimized combination, not available in nature, of two or more responses to specific excitation" by being fully programmable. That is, unlike in a conventional metamaterial, the relationship between a specific excitation and response is governed by sensing, actuation, and a computer program that implements the desired logic.

The idea of creating materials that embed computation is closely related to the concept of programmable matter, a term coined in 1991 by Toffoli and Margolus, describing dense arrays of computing elements that could solve complex finite-element like simulations of material systems, and then later developed to describe a class of materials consisting of identical, mobile building blocks, also known as catoms that are fully reconfigurable, therefore allowing materials to arbitrarily change their physical properties.

Robotic materials build up on the original concept of programmable matter, but focus on the structural properties of the embedding polymers without claim of universal property changes. Here the term "robotic" refers to the confluence of sensing, actuation, and computation.

## Applications

Robotic materials allow to off-load computation inside the material, most notably signal processing that arises during high-bandwidth sensing applications or feedback control that is required by fine-grained distributed actuation. Examples for such applications include camouflage, shape change, load balancing, and robotic skins as well as equipping robots with more autonomy by off-loading some of the signal processing and controls into the material.

# THREE LAWS OF ROBOTICS

The Three Laws of Robotics (often shortened to The Three Laws or known as Asimov's Laws) are a set of rules devised by the science fiction author Isaac Asimov. The rules were introduced in his 1942 short story "Runaround" (included in the 1950 collection I, Robot), although they had been foreshadowed in a few earlier stories. The Three Laws, quoted as being from the "Handbook of Robotics, 56th Edition, 2058 A.D.", are:

### First Law

A robot may not injure a human being or, through inaction, allow a human being to come to harm.

### Second Law

A robot must obey the orders given it by human beings except where such orders would conflict with the First Law.

### Third Law

A robot must protect its own existence as long as such protection does not conflict with the First or Second Laws.

These form an organizing principle and unifying theme for Asimov's robotic-based fiction, appearing in his *Robot* series, the stories linked to it, and his *Lucky Starr* series of young-adult fiction. The Laws are incorporated into almost all of the positronic robots appearing in his fiction, and cannot be bypassed, being intended as a safety feature. Many of Asimov's robot-focused stories involve robots behaving in unusual and counter-intuitive ways as an unintended consequence of how the robot applies the Three Laws to the situation in which it finds itself. Other authors working in Asimov's fictional universe have adopted them and references, often parodic, appear throughout science fiction as well as in other genres.

The original laws have been altered and elaborated on by Asimov and other authors. Asimov himself made slight modifications to the first three in various books and short stories to further develop how robots would interact with humans and each other. In later fiction where robots had taken responsibility for government of whole planets and human civilizations, Asimov also added a fourth, or zeroth law, to precede the others:

## Zeroth Law

A robot may not harm humanity, or, by inaction, allow humanity to come to harm.

The Three Laws, and the zeroth, have pervaded science fiction and are referred to in many books, films, and other media. They have impacted thought on ethics of artificial intelligence as well.

## Alterations

### By Asimov

Asimov's stories test his Three Laws in a wide variety of circumstances leading to proposals and rejection of modifications. Science fiction scholar James Gunn writes in 1982, "The Asimov robot stories as a whole may respond best to an analysis on this basis: the ambiguity in the Three Laws and the ways in which Asimov played twenty-nine variations upon a theme". While the original set of Laws provided inspirations for many stories, Asimov introduced modified versions from time to time.

### First Law Modified

In "Little Lost Robot" several NS-2, or "Nestor", robots are created with only part of the First Law. It reads:

- A robot may not harm a human being.

This modification is motivated by a practical difficulty as robots have to work alongside human beings who are exposed to low doses of radiation. Because their positronic brains are highly sensitive to gamma rays the robots are rendered inoperable by doses reasonably safe for humans. The robots are being destroyed attempting to rescue the humans who are in no actual danger but "might forget to leave" the irradiated area within the exposure time limit. Removing the First Law's "inaction" clause solves this problem but creates the possibility of an even greater one: a robot could initiate an action that would harm a human (dropping a heavy weight and failing to catch it is the example given in the text), knowing that it was capable of preventing the harm and then decide not to do so.

Gaia is a planet with collective intelligence in the *Foundation* series which adopts a law similar to the First Law, and the Zeroth Law, as its philosophy:

Gaia may not harm life or allow life to come to harm.

## Zeroth Law Added

Asimov once added a "Zeroth Law"—so named to continue the pattern where lower-numbered laws supersede the higher-numbered laws—stating that a robot must not harm humanity. The robotic character R. Daneel Olivaw was the first to give the Zeroth Law a name in the novel *Robots and Empire*; however, the character Susan Calvin articulates the concept in the short story "The Evitable Conflict".

In the final scenes of the novel *Robots and Empire*, R. Giskard Reventlov is the first robot to act according to the Zeroth Law. Giskard is telepathic, like the robot Herbie in the short story "Liar!", and tries to apply the Zeroth Law through his understanding of a more subtle concept of "harm" than most robots can grasp. However, unlike Herbie, Giskard grasps the philosophical concept of the Zeroth Law allowing him to harm individual human beings if he can do so in service to the abstract concept of humanity. The Zeroth Law is never programmed into Giskard's brain but instead is a rule he attempts to comprehend through pure metacognition. Though he fails – it ultimately destroys his positronic brain as he is not certain whether his choice will turn out to be for the ultimate good of humanity or not – he gives his successor R. Daneel Olivaw his telepathic abilities. Over the course of many thousands of years Daneel adapts himself to be able to fully obey the Zeroth Law. As Daneel formulates it, in the novels *Foundation and Earth* and *Prelude to Foundation*, the Zeroth Law reads:

A robot may not harm humanity, or, by inaction, allow humanity to come to harm.

A condition stating that the Zeroth Law must not be broken was added to the original Three Laws, although Asimov recognized the difficulty such a law would pose in practice. Asimov's novel *Foundation and Earth* contains the following passage:

Trevize frowned. "How do you decide what is injurious, or not injurious, to humanity as a whole?"

"Precisely, sir," said Daneel. "In theory, the Zeroth Law was the answer to our problems. In practice, we could never decide. A human being is a concrete object. Injury to a person can be estimated and judged. Humanity is an abstraction."

A translator incorporated the concept of the Zeroth Law into one of Asimov's novels before Asimov himself made the law explicit. Near the climax of *The Caves of Steel*, Elijah Baley makes a bitter comment to himself thinking that the First Law forbids a robot from harming a human being. He determines that it must be so unless the robot is clever enough to comprehend that its actions are for humankind's long-term good. In Jacques Brécard's 1956 French translation entitled *Les Cavernes d'acier* Baley's thoughts emerge in a slightly different way:

A robot may not harm a human being, unless he finds a way to prove that ultimately the harm done would benefit humanity in general.

## Removal of the Three Laws

Three times during his writing career, Asimov portrayed robots that disregard the Three Laws entirely. The first case was a short-short storyentitled "First Law" and is often considered an insignificant "tall tale" or even apocryphal. On the other hand, the short story "Cal" (from the collection *Gold*), told by a first-person robot narrator, features a robot who disregards the Three Laws because he has found something far more important—he wants to be a writer. Humorous, partly autobiographical and unusually experimental in style, "Cal" has been regarded as one of *Gold's* strongest stories. The third is a short story entitled "Sally" in which cars fitted with positronic brains are apparently able to harm and kill humans in disregard of the First Law. However, aside from the positronic brain concept, this story does not refer to other robot stories and may not be set in the same continuity.

The title story of the *Robot Dreams* collection portrays LVX-1, or "Elvex", a robot who enters a state of unconsciousness and dreams thanks to the unusual fractal construction of his positronic brain. In his dream the first two Laws are absent and the Third Law reads "A robot must protect its own existence".

Asimov took varying positions on whether the Laws were optional: although in his first writings they were simply carefully engineered safeguards, in later stories Asimov stated that they were an inalienable part of the mathematical foundation underlying the positronic brain. Without the basic theory of the Three Laws the fictional scientists of Asimov's universe would be unable to design a workable brain unit. This is historically consistent: the occasions where roboticists modify the Laws generally occur early within the stories' chronology and at a time when there is less existing work to be re-done. In "Little Lost Robot" Susan Calvin considers modifying the Laws to be a terrible idea, although possible, while centuries later Dr. Gerrigel in *The Caves of Steel* believes it to be impossible.

The character Dr. Gerrigel uses the term "Asenion" to describe robots programmed with the Three Laws. The robots in Asimov's stories, being Asenion robots, are incapable of knowingly violating the Three Laws but, in principle, a robot in science fiction or in the real world could be non-Asenion. "Asenion" is a misspelling of the name Asimov which was made by an editor of the magazine *Planet Stories*.

Asimov used this obscure variation to insert himself into *The Caves of Steel* just like he referred to himself as "Azimuth or, possibly, Asymptote" in *Thiotimoline to the Stars*, in much the same way that Vladimir Nabokov appeared in *Lolita* anagrammatically disguised as "Vivian Darkbloom".

Characters within the stories often point out that the Three Laws, as they exist in a robot's mind, are not the written versions usually quoted by humans but abstract mathematical concepts upon which a robot's entire developing consciousness is based. This concept is largely fuzzy and unclear in earlier stories depicting very rudimentary robots who are only programmed to comprehend basic physical tasks, where the Three Laws act as an overarching safeguard, but by the era of *The Caves of Steel* featuring robots with human or beyond-human intelligence the Three Laws have become the underlying basic ethical worldview that determines the actions of all robots.

# ASPECTS OF ROBOTICS

## Robotic Aspects

Mechanical construction.

Electrical aspect.

```
// Dev-C++ 4.9.9.2
// Project Type: Win32 GUI
// Window: Window Header
#include <Windows.h>
#include "resource.h"
// Window: Window Name
#ifdef NULL
#undef NULL
#define NULL 0
#endif
#define Wnd_Class "WIN_CHK"
#define Wnd_Title "預設視窗"
// Window: Window Parameters
static UINT WndPos_X = 0, WndPos_Y = 0;
// 400 x 300
static UINT WndPos_Width = 400, WndPos_Height = 300;
static HWND hwndWnd = 0;
static HINSTANCE hinstWnd = 0;
LRESULT CALLBACK WndProc(HWND, UINT, WPARAM, LPARAM);
BOOL ProcMsg(void);
BOOL BuildWnd( const char*, const char*);
void InitWindow_PositionCenter( UINT&, UINT&, UINT, UINT, BOOL);
// Window: Window Entry
int WINAPI WinMain(HINSTANCE hInstance, HINSTANCE hPrevInstance,
                   LPSTR lpCmdLine, int nShowCmd)
{
    //==START of WinMain==//
    if ( (hwndWnd = ::FindWindow( Wnd_Class, Wnd_Title)) != NULL )
    {
        ::SetForegroundWindow(hwndWnd);
        return NULL;
    }
    if ( BuildWnd( Wnd_Class, Wnd_Title) == TRUE )
    {
        while ( ProcMsg() == TRUE );
    }
    //==END of WinMain==//
    return NULL;
}
```

A level of programming.

There are many types of robots; they are used in many different environments and for many different uses, although being very diverse in application and form they all share three basic similarities when it comes to their construction:

1.  Robots all have some kind of mechanical construction, a frame, form or shape designed to achieve a particular task. For example, a robot designed to travel across heavy dirt or mud, might use caterpillar tracks. The mechanical aspect is mostly the creator's solution to completing the assigned task and dealing with the physics of the environment around it. Form follows function.

2.  Robots have electrical components which power and control the machinery. For example,

the robot with caterpillar tracks would need some kind of power to move the tracker treads. That power comes in the form of electricity, which will have to travel through a wire and originate from a battery, a basic electrical circuit. Even petrol powered machines that get their power mainly from petrol still require an electric current to start the combustion process which is why most petrol powered machines like cars, have batteries. The electrical aspect of robots is used for movement (through motors), sensing (where electrical signals are used to measure things like heat, sound, position, and energy status) and operation (robots need some level of electrical energy supplied to their motors and sensors in order to activate and perform basic operations).

3. All robots contain some level of computer programming code. A program is how a robot decides when or how to do something. In the caterpillar track example, a robot that needs to move across a muddy road may have the correct mechanical construction and receive the correct amount of power from its battery, but would not go anywhere without a program telling it to move. Programs are the core essence of a robot, it could have excellent mechanical and electrical construction, but if its program is poorly constructed its performance will be very poor (or it may not perform at all). There are three different types of robotic programs: remote control, artificial intelligence and hybrid. A robot with remote control programing has a preexisting set of commands that it will only perform if and when it receives a signal from a control source, typically a human being with a remote control. It is perhaps more appropriate to view devices controlled primarily by human commands as falling in the discipline of automation rather than robotics. Robots that use artificial intelligence interact with their environment on their own without a control source, and can determine reactions to objects and problems they encounter using their preexisting programming. Hybrid is a form of programming that incorporates both AI and RC functions.

# ROBOT KINEMATICS

Robot kinematics applies geometry to the study of the movement of multi-degree of freedom kinematic chains that form the structure of robotic systems. The emphasis on geometry means that the links of the robot are modeled as rigid bodies and its joints are assumed to provide pure rotation or translation.

Robot kinematics studies the relationship between the dimensions and connectivity of kinematic chains and the position, velocity and acceleration of each of the links in the robotic system, in order to plan and control movement and to compute actuator forces and torques. The relationship between mass and inertia properties, motion, and the associated forces and torques is studied as part of robot dynamics.

## Kinematic Equations

A fundamental tool in robot kinematics is the kinematics equations of the kinematic chains that form the robot. These non-linear equations are used to map the joint parameters to the configuration of the robot system. Kinematics equations are also used in biomechanics of the skeleton and computer animation of articulated characters.

Forward kinematics uses the kinematic equations of a robot to compute the position of the end-effector from specified values for the joint parameters. The reverse process that computes the joint parameters that achieve a specified position of the end-effector is known as inverse kinematics. The dimensions of the robot and its kinematics equations define the volume of space reachable by the robot, known as its workspace.

There are two broad classes of robots and associated kinematics equations: serial manipulators and parallel manipulators. Other types of systems with specialized kinematics equations are air, land, and submersible mobile robots, hyper-redundant, or snake, robots and humanoid robots.

## Forward Kinematics

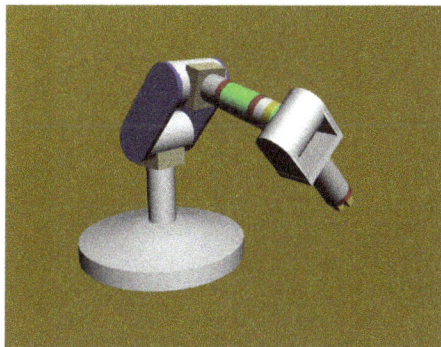

An articulated six DOF robotic arm uses forward kinematics to position the gripper.

The forward kinematics equations define the trajectory of the end-effector of a PUMA robot reaching for parts.

Forward kinematics refers to the use of the kinematic equations of a robot to compute the position of the end-effector from specified values for the joint parameters.

The kinematics equations of the robot are used in robotics, computer games, and animation. The reverse process that computes the joint parameters that achieve a specified position of the end-effector is known as inverse kinematics.

The kinematics equations for the series chain of a robot are obtained using a rigid transformation [Z] to characterize the relative movement allowed at each joint and separate rigid transformation [X] to define the dimensions of each link. The result is a sequence of rigid transformations alternating joint and link transformations from the base of the chain to its end link, which is equated to the specified position for the end link,

$$[T] = [Z_1][X_1][Z_2][X_2]...[X_{n-1}][Z_n],$$

where [T] is the transformation locating the end-link. These equations are called the kinematics equations of the serial chain.

## Link Transformations

In 1955, Jacques Denavit and Richard Hartenberg introduced a convention for the definition of the joint matrices [Z] and link matrices [X] to standardize the coordinate frame for spatial linkages. This convention positions the joint frame so that it consists of a screw displacement along the Z-axis,

$$[Z_i] = \text{Trans}_{Z_i}(d_i)\text{Rot}_{Z_i}(\theta_i),$$

and it positions the link frame so it consists of a screw displacement along the X-axis,

$$[X_i] = \text{Trans}_{X_i}(a_{i,i+1})\text{Rot}_{X_i}(\alpha_{i,i+1}).$$

Using this notation, each transformation-link goes along a serial chain robot, and can be described by the coordinate transformation,

$$^{i-1}T_i = [Z_i][X_i] = \text{Trans}_{Z_i}(d_i)\text{Rot}_{Z_i}(\theta_i)\text{Trans}_{X_i}(a_{i,i+1})\text{Rot}_{X_i}(\alpha_{i,i+1}),$$

where $\theta i$, $di$, $ai,i+1$ and $ai,i+1$ are known as the Denavit-Hartenberg parameters.

## Kinematics Equations Revisited

The kinematics equations of a serial chain of $n$ links, with joint parameters $\theta i$ are given by,

$$[T] = {}^0T_n = \prod_{i=1}^{n} {}^{i-1}T_i(\theta_i),$$

where $^{i-1}T_i(\theta_i)$ is the transformation matrix from the frame of link $i$ to link $i-1$. In robotics, these are conventionally described by Denavit–Hartenberg parameters.

## Denavit-Hartenberg Matrix

The matrices associated with these operations are:

$$\text{Trans}_{Z_i}(d_i) = \begin{bmatrix} 1 & 0 & 0 & 0 \\ 0 & 1 & 0 & 0 \\ 0 & 0 & 1 & d_i \\ 0 & 0 & 0 & 1 \end{bmatrix}, \quad \text{Rot}_{Z_i}(\theta_i) = \begin{bmatrix} \cos\theta_i & -\sin\theta_i & 0 & 0 \\ \sin\theta_i & \cos\theta_i & 0 & 0 \\ 0 & 0 & 1 & 0 \\ 0 & 0 & 0 & 1 \end{bmatrix}.$$

Similarly,

$$\mathrm{Trans}_{X_i}(a_{i,i+1}) = \begin{bmatrix} 1 & 0 & 0 & a_{i,i+1} \\ 0 & 1 & 0 & 0 \\ 0 & 0 & 1 & 0 \\ 0 & 0 & 0 & 1 \end{bmatrix}, \quad \mathrm{Rot}_{X_i}(\alpha_{i,i+1}) = \begin{bmatrix} 1 & 0 & 0 & 0 \\ 0 & \cos\alpha_{i,i+1} & -\sin\alpha_{i,i+1} & 0 \\ 0 & \sin\alpha_{i,i+1} & \cos\alpha_{i,i+1} & 0 \\ 0 & 0 & 0 & 1 \end{bmatrix}.$$

The use of the Denavit-Hartenberg convention yields the link transformation matrix, $[i\text{-}1Ti]$ as

$$^{i-1}T_i = \begin{bmatrix} \cos\theta_i & -\sin\theta_i\cos\alpha_{i,i+1} & \sin\theta_i\sin\alpha_{i,i+1} & \alpha_{i,i+1}\cos\theta_i \\ \sin\theta_i & \cos\theta_i\cos\alpha_{i,i+1} & -\cos\theta_i\sin\alpha_{i,i+1} & \alpha_{i,i+1}\sin\theta_i \\ 0 & \sin\alpha_{i,i+1} & \cos\alpha_{i,i+1} & d_i \\ 0 & 0 & 0 & 1 \end{bmatrix}$$

known as the *Denavit-Hartenberg matrix*.

## Inverse Kinematics

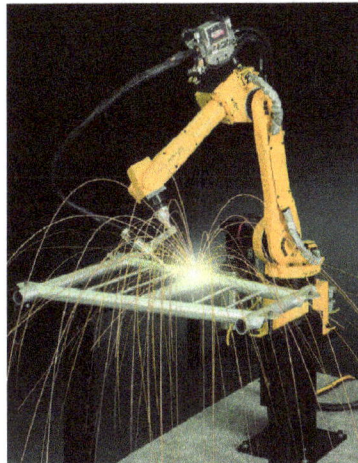

An industrial robot performing arc welding. Inverse kinematics computes
the joint trajectories needed for the robot to guide the welding tip along the part.

Inverse kinematics is the mathematical process of recovering the movements of an object in the world from some other data, such as a film of those movements, or a film of the world as seen by a camera which is itself making those movements. This is useful in robotics and in film animation.

In robotics, inverse kinematics makes use of the kinematics equations to determine the joint parameters that provide a desired position for each of the robot's end-effectors. Specification of the movement of a robot so that its end-effectors achieve the desired tasks is known as motion planning. Inverse kinematics transforms the motion plan into joint actuator trajectories for the robot. Similar formulae determine the positions of the skeleton of an animated character that is to move in a particular way in a film, or of a vehicle such as a car or boat containing the camera which is shooting a scene of a film. Once a vehicle's motions are known, they can be used to determine the constantly-changing viewpoint for computer-generated imagery of objects in the landscape such as buildings, so that these objects change in perspective while themselves not appearing to move as the vehicle-borne camera goes past them.

The movement of a kinematic chain, whether it is a robot or an animated character, is modeled by the kinematics equations of the chain. These equations define the configuration of the chain in terms of its joint parameters. Forward kinematics uses the joint parameters to compute the configuration of the chain, and inverse kinematics reverses this calculation to determine the joint parameters that achieve a desired configuration.

## Kinematic Analysis

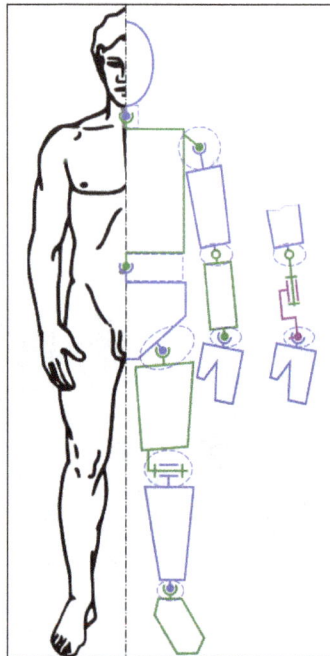

A model of the human skeleton as a kinematic chain allows positioning using inverse kinematics.

Kinematic analysis is one of the first steps in the design of most industrial robots. Kinematic analysis allows the designer to obtain information on the position of each component within the mechanical system. This information is necessary for subsequent dynamic analysis along with control paths.

Inverse kinematics is an example of the kinematic analysis of a constrained system of rigid bodies, or kinematic chain. The kinematic equations of a robot can be used to define the loop equations of a complex articulated system. These loop equations are non-linear constraints on the configuration parameters of the system. The independent parameters in these equations are known as the degrees of freedom of the system.

While analytical solutions to the inverse kinematics problem exist for a wide range of kinematic chains, computer modeling and animation tools often use Newton's method to solve the non-linear kinematics equations.

Other applications of inverse kinematic algorithms include interactive manipulation, animation control and collision avoidance.

# ARTIFICIAL INTELLIGENCE

Artificial Intelligence (AI) is arguably the most exciting field in robotics. It's certainly the most controversial: Everybody agrees that a robot can work in an assembly line, but there's no consensus on whether a robot can ever be intelligent.

Like the term "robot" itself, artificial intelligence is hard to define. Ultimate AI would be a recreation of the human thought process -- a man-made machine with our intellectual abilities. This would include the ability to learn just about anything, the ability to reason, the ability to use language and the ability to formulate original ideas. Roboticists are nowhere near achieving this level of artificial intelligence, but they have made a lot of progress with more limited AI. Today's AI machines can replicate some specific elements of intellectual ability.

Computers can already solve problems in limited realms. The basic idea of AI problem-solving is very simple, though its execution is complicated. First, the AI robot or computer gathers facts about a situation through sensors or human input. The computer compares this information to stored data and decides what the information signifies. The computer runs through various possible actions and predicts which action will be most successful based on the collected information. Of course, the computer can only solve problems it's programmed to solve -- it doesn't have any generalized analytical ability. Chess computers are one example of this sort of machine.

Some modern robots also have the ability to learn in a limited capacity. Learning robots recognize if a certain action (moving its legs in a certain way, for instance) achieved a desired result (navigating an obstacle). The robot stores this information and attempts the successful action the next time it encounters the same situation. Again, modern computers can only do this in very limited situations. They can't absorb any sort of information like a human can. Some robots can learn by mimicking human actions. In Japan, roboticists have taught a robot to dance by demonstrating the moves themselves.

The real challenge of AI is to understand how natural intelligence works. Developing AI isn't like building an artificial heart -- scientists don't have a simple, concrete model to work from. We do know that the brain contains billions and billions of neurons, and that we think and learn by establishing electrical connections between different neurons. But we don't know exactly how all of these connections add up to higher reasoning, or even low-level operations. The complex circuitry seems incomprehensible.

Because of this, AI research is largely theoretical. Scientists hypothesize on how and why we learn and think, and they experiment with their ideas using robots. Brooks and his team focus on humanoid robots because they feel that being able to experience the world like a human is essential to developing human-like intelligence. It also makes it easier for people to interact with the robots, which potentially makes it easier for the robot to learn.

Just as physical robotic design is a handy tool for understanding animal and human anatomy, AI research is useful for understanding how natural intelligence works. For some roboticists, this insight is the ultimate goal of designing robots. Others envision a world where we live side by side with intelligent machines and use a variety of lesser robots for manual labor, health care and communication. A number of robotics experts predict that robotic evolution will ultimately turn us into cyborgs -- humans integrated with machines. Conceivably, people in the future could load their minds into a sturdy robot and live for thousands of years.

Commercially available applications include the use of AI to:

- Enable robots to sense and respond to their environment: This vastly increases the range of functions robots can perform.

- Optimise robot and process performance, saving companies money.

- Enable robots to function as mobile, interactive information systems in numerous settings from public spaces to hospitals to retail outlets, saving individuals time.

## Examples

### Sense-and-respond

Identifying, picking, and passing objects: Traditionally, robots have been able to pick up objects in a pre-programmed trajectory in which the object must be known and in the expected place. Robots equipped with sensors can now be programmed using artificial intelligence to identify specific objects regardless of their spatial location. 3D vision software allows the robot to detect objects that are hidden by other objects. Through machine learning, one of the technologies classed as AI, the robot can teach itself in a very short time how to pick up an object it has not encountered before, applying the appropriate level of force. The machine learning algorithm continues to improve as it picks. Picking technology is advancing rapidly but it is currently very difficult for robots to pick objects that are not rigid – for example, goods in plastic wrapping or floppy materials, or have irregular and variable shapes – such as fruit and vegetables – with an accuracy and speed that is commercially viable.

Inspection: Artificial intelligence enables robots to inspect a wide variety of objects to detect faults – from fruit and vegetables to underwater pipelines.

Mobility: AI technologies are enabling advanced mobility in robots. Whilst robots have been mobile for over 60 years (the first Automated Guided Vehicle was introduced in 1953), AI enables robot mobility in unpredictable environments. Mobile robots have traditionally been programmed to execute a specific set of manoeuvres in a linear fashion, guided by signals (magnetic, laser, lidar) from devices installed for this purpose in their environment. They have not traditionally been programmed to deal with unexpected events – for example, if they encounter an obstacle, they can stop to avoid collision, but they will not be able to find an alternative route to their goal. In contrast, an AI-enabled mobile robot gets from A to B by building a real-time map (or updating a preprogrammed map in real-time) of its environment and of its location within that environment, planning a path to the programmed goal, sensing obstacles and re-planning a path in-situ. Mobile robots using AI are in commercial use in a number of industries and applications such as:

- Fetching and carrying goods in factories, warehouses, hospitals.

- Performing inventory management (mobile robots using RFID scanners or vision technologies).

- Cleaning – from offices to large pieces of equipment such as ship hulls.

- Exploration of environments dangerous for humans – e.g deep-sea, space, contaminated environments.

## Process Optimisation

AI is used to optimise robot accuracy and reliability. Most large industrial robot manufacturers offer customers services using AI to analyse data from robots in real time to predict whether and when a robot is likely to require maintenance, enabling manufacturers to avoid costly machine downtime. Robot performance can also be optimised through analysis of data from sensors - tracking, for example, its movement and power consumption. The robot programme can be adjusted automatically based on the output of the AI algorithm.

Predictive maintenance and process optimisation do not require AI. However, AI technologies improve the speed and accuracy of both activities, resulting in cost savings. In large-scale manufacturing automation projects robots are typically connected to other machinery – including other robots – and AI is used to optimise the whole process, analysing data from all machines.

## Mobile Information Robots

Mobile robots are being used as information booths to assist customers in environments such as hotels, hospitals, airports and shops. They can answer questions, lead customers to requested products or locations and can video-link the customer to a human service agent.

# FUTURE OF ROBOTICS

Practically no fantast and even no ordinary realist would risk describing the future society without robots, in particular, androids. It is clear that now one can see the prototypes which show the achievements of the scientists and engineers in this direction. Though there are still a number of unresolved technical tasks, it is possible to say surely that in the next 20 years, there will be more perfect and cheaper technologies in this area, which will lead to the formation of the market of robots (androids) of the most different functional purposes and complexity levels. It means that androids (and other robots) will live and work among people, entertaining and helping them with their daily physical and intellectual work.

Android is not a simple robot similar to a person, it is the result of the achievements of the whole direction in robotics which seeks to create robots absolutely similar to a person. The highest achievement of robotics will be the android which will not practically distinguish from an ordinary person.

One of the most basic difficulties in the way to the achievement of this purpose is an imitation of the natural communicative behavior of the person. The person not only talks but expresses the most different emotions and thoughts. To teach the robot to understand the speech, saying to it, and express certain feelings is practically impossible to fulfill at this stage. In the future, robots will learn to understand not only the usual speech but also nonverbal signals (for example, gestures and mimicry). They will be able to communicate with people on different topics.

The requirement of an external similarity of the android with the person results in the need of the solution of a number of technical tasks. For example, straight moving of the android has to be steady, and his "hands" have to allow it to move freights and to manipulate objects.

The movements of the android, most likely, will remain not "natural", while his hands and legs will work with the help of the mechanical devices and systems, which have a number of shortcomings. The decisions on the basis of special synthetic materials, which could assume the function of muscles, would be very useful and would replace electric and pneumatic devices. Nowadays, synthetic muscles are being developed.

By 2020, there will have been a large number of almost completely robotized productions. By 2015-2020, robots will have been actively applied in agriculture. By this time there will have been robots in the streets of the cities. Those will be robots loaders and robots cleaners. The most part of the land transport will have been automated by 2020-2030. Cars "will grow wiser" and in some time, they will take all the process of driving under control (Nusca).

Robots will find their application in medicine. In some areas of medicine, they will be capable of working more effectively than people. By 2020-2025, the most part of performed medical operations will have been carried out by robots. At this time, there will have to be first micro robots which watch over the health of the person in his/her body.

In the future, robots will be able to move not only on the ground but in all imaginable ways. Robots will be capable of floating in the rivers, seas, and ocean depths, and soaring in the air. Thanks to their evolutionary algorithms of creation, many robots will be able to change their form and also structure, depending on the current situation.

According to Bill Gates, in the near future, "robotics is expected to have revolutionary changes which marked the break in computer facilities more than 30 years ago".

## References

- Robotics-introduction: geeksforgeeks.org,  Retrieved 1 January, 2019

- Neurorobotics: scholarpedia.org,  Retrieved 2 February, 2019

- Pearce, Joshua M. (2014-01-01). "Introduction to Open-Source Hardware for Science". Chapter 1 - Introduction to Open-Source Hardware for Science. Boston: Elsevier. Pp. 1–11. Doi:10.1016/b978-0-12-410462-4.00001-9. ISBN 9780124104624

- Robot6: science.howstuffworks.com,  Retrieved 3 March, 2019

- "Robotics: About the Exhibition". The Tech Museum of Innovation. Archived from the original on 2008-09-13. Retrieved 2008-09-15

- Media-Backgrounder-on-Artificial-Intelligence-in-Robotics-May: ifr.org,  Retrieved 4 April, 2019

- Asada, M.; Hosoda, K.; Kuniyoshi, Y.; Ishiguro, H.; Inui, T.; Yoshikawa, Y.; Ogino, M.; Yoshida, C. (2009). "Cognitive developmental robotics: a survey". IEEE Transactions on Autonomous Mental Development. 1 (1): 12–34. Doi:10.1109/tamd.2009.2021702

- Future-of-robotics, Research, essays: essayswriters.com, Retrieved 5 May, 2019

- Jenkins, John H. (2002). "Review of "Cal"". Jenkins' Spoiler-Laden Guide to Isaac Asimov. Archived from the original on 2009-09-11.  Retrieved 2009-06-26

# 2

# Robots and its Types

A robot is a machine that is capable of carrying out complex actions automatically. Some of the different types of robots are aerobot, agricultural robot, autonomous robots, mobile robots, legged robots, humanoid robots, medical robots, entertainment robots, etc. This chapter discusses these types of robots in detail.

## ROBOTS

A robot is a machine designed to execute one or more tasks automatically with speed and precision. There are as many different types of robots as there are tasks for them to perform.

Robots that resemble humans are known as androids; however, many robots aren't built on the human model. Industrial robots, for example, are often designed to perform repetitive tasks that aren't facilitated by a human-like construction. A robot can be remotely controlled by a human operator, sometimes from a great distance. A telechir is a complex robot that is remotely controlled by a human operator for a telepresence system, which gives that individual the sense of being on location in a remote, dangerous or alien environment and the ability to interact with it. Telepresence robots, which simulate the experience and some of the capabilities of being physically present, can enable remote business consultations, healthcare, home monitoring and childcare, among many other possibilities.

An autonomous robot acts as a stand-alone system, complete with its own computer (called the controller). The most advanced example is the smart robot, which has a built-in artificial intelligence (AI) system that can learn from its environment and its experience and build on its capabilities based on that knowledge.

Swarm robots, sometimes referred to as insect robots, work in fleets ranging in number from a few to thousands, with all fleet members under the supervision of a single controller. The term arises from the similarity of the system to a colony of insects, where the individuals and behaviors are simple but the fleet as a whole can be sophisticated.

Robots are sometimes grouped according to the time frame in which they were first widely used. First-generation robots date from the 1970s and consist of stationary, nonprogrammable, electromechanical devices without sensors. Second-generation robots were developed in the 1980s and can contain sensors and programmable controllers. Third-generation robots were developed between approximately 1990 and the present. These machines can be stationary or mobile, autonomous or insect type, with sophisticated programming, speech recognition and/or synthesis, and other advanced features. Fourth-generation robots are in the research-and-development phase, and include features such as artificial intelligence, self-replication, self-assembly, and nanoscale size (physical dimensions on the order of nanometers, or units of 10-9 meter).

Some advanced robots are called androids because of their superficial resemblance to human beings. Androids are mobile, usually moving around on wheels or a track drive (robots legs are unstable and difficult to engineer). The android is not necessarily the end point of robot evolution. Some of the most esoteric and powerful robots do not look or behave anything like humans. The ultimate in robotic intelligence and sophistication might take on forms yet to be imagined.

## Robots are All Around us

Using this Working Definition of a Robot, take a Quick Look at the Robots in common use:

- Industrial: Robots were quickly put to use in industry, beginning with Unimate, a robot designed by George Devol in 1959 for General Motors. Considered to be the first industrial robot, Ultimate was a robotic arm used to manipulate hot die-cast parts in automobile manufacturing, a task that was dangerous for humans to perform.

- Medical: Robots in medicine perform a wide range of tasks, including performing surgery, assisting in rehabilitation, or automatically disinfecting hospital rooms and surgical suites.

- Consumer: Perhaps the best-recognized household robot is the Roomba vacuum cleaner, which automatically cleans the floors around your house. Along the same line are a number of robotic lawn mowers that keep your grass clipped for you.

- Robots you didn't know were robots: The list contains robots you come across every day, but probably don't think of as robots: automatic car washes, speeding or red light cameras, automatic door openers, elevators, and some kitchen appliances.

## Uses of Robots

At present, there are two main types of robots, based on their use: general-purpose autonomous robots and dedicated robots.

A general-purpose robot acts as a guide during the day and a security guard at night.

Robots can be classified by their specificity of purpose. A robot might be designed to perform one particular task extremely well, or a range of tasks less well. All robots by their nature can be re-programmed to behave differently, but some are limited by their physical form. For example, a factory robot arm can perform jobs such as cutting, welding, gluing, or acting as a fairground ride, while a pick-and-place robot can only populate printed circuit boards.

## General-purpose Autonomous Robots

General-purpose autonomous robots can perform a variety of functions independently. General-purpose autonomous robots typically can navigate independently in known spaces, handle their own re-charging needs, interface with electronic doors and elevators and perform other basic tasks. Like computers, general-purpose robots can link with networks, software and accessories that increase their usefulness. They may recognize people or objects, talk, provide companionship, monitor environmental quality, respond to alarms, pick up supplies and perform other useful tasks. General-purpose robots may perform a variety of functions simultaneously or they may take on different roles at different times of day. Some such robots try to mimic human beings and may even resemble people in appearance; this type of robot is called a humanoid robot. Humanoid robots are still in a very limited stage, as no humanoid robot can, as of yet, actually navigate around a room that it has never been in. Thus, humanoid robots are really quite limited, despite their intelligent behaviors in their well-known environments.

## Factory Robots

## Car Production

Over the last three decades, automobile factories have become dominated by robots. A typical factory contains hundreds of industrial robotsworking on fully automated production lines, with one robot for every ten human workers. On an automated production line, a vehicle chassis on a conveyor is welded, glued, painted and finally assembled at a sequence of robot stations.

## Packaging

Industrial robots are also used extensively for palletizing and packaging of manufactured goods, for example for rapidly taking drink cartons from the end of a conveyor belt and placing them into boxes, or for loading and unloading machining centers.

## Electronics

Mass-produced printed circuit boards (PCBs) are almost exclusively manufactured by pick-and-place robots, typically with SCARAmanipulators, which remove tiny electronic components from strips or trays, and place them on to PCBs with great accuracy. Such robots can place hundreds of thousands of components per hour, far out-performing a human in speed, accuracy, and reliability.

## Automated Guided Vehicles (AGVs)

Mobile robots, following markers or wires in the floor, or using vision or lasers, are used to transport goods around large facilities, such as warehouses, container ports, or hospitals.

An intelligent AGV drops-off goods without needing lines or beacons in the workspace.

### Early AGV-style Robots

Limited to tasks that could be accurately defined and had to be performed the same way every time. Very little feedback or intelligence was required, and the robots needed only the most basic exteroceptors(sensors). The limitations of these AGVs are that their paths are not easily altered and they cannot alter their paths if obstacles block them. If one AGV breaks down, it may stop the entire operation.

### Interim AGV Technologies

Developed to deploy triangulation from beacons or bar code grids for scanning on the floor or ceiling. In most factories, triangulation systems tend to require moderate to high maintenance, such as daily cleaning of all beacons or bar codes. Also, if a tall pallet or large vehicle blocks beacons or a bar code is marred, AGVs may become lost. Often such AGVs are designed to be used in human-free environments.

### Intelligent AGVs (i-AGVs)

Such as SmartLoader, SpeciMinder, ADAM, Tug Eskorta, and MT 400 with Motivity are designed for people-friendly workspaces. They navigate by recognizing natural features. 3D scanners or other means of sensing the environment in two or three dimensions help to eliminate cumulative errors in dead-reckoning calculations of the AGV's current position. Some AGVs can create maps of their environment using scanning lasers with simultaneous localization and mapping (SLAM) and use those maps to navigate in real time with other path planning and obstacle avoidance algorithms. They are able to operate in complex environments and perform non-repetitive and non-sequential tasks such as transporting photomasks in a semiconductor lab, specimens in hospitals and goods in warehouses. For dynamic areas, such as warehouses full of pallets, AGVs require additional strategies using three-dimensional sensors such as time-of-flight or stereovision cameras.

## Dirty, Dangerous, Dull or Inaccessible Tasks

There are many jobs which humans would rather leave to robots. The job may be boring, such as domestic cleaning or sports field line marking, or dangerous, such as exploring inside a volcano. Other jobs are physically inaccessible, such as exploring another planet, cleaning the inside of a long pipe, or performing laparoscopic surgery.

## Space Probes

Almost every unmanned space probe ever launched was a robot. Some were launched in the 1960s with very limited abilities, but their ability to fly and land (in the case of Luna 9) is an indication of their status as a robot. This includes the Voyager probes and the Galileo probes, among others.

## Telerobots

A U.S. Marine Corps technician prepares to use a telerobot to detonate a buried improvised explosive devicenear Camp Fallujah, Iraq.

Teleoperated robots, or telerobots, are devices remotely operated from a distance by a human operator rather than following a predetermined sequence of movements, but which has semi-autonomous behaviour. They are used when a human cannot be present on site to perform a job because it is dangerous, far away, or inaccessible. The robot may be in another room or another country, or may be on a very different scale to the operator. For instance, a laparoscopic surgery robot allows the surgeon to work inside a human patient on a relatively small scale compared to open surgery, significantly shortening recovery time. They can also be used to avoid exposing workers to the hazardous and tight spaces such as in duct cleaning. When disabling a bomb, the operator sends a small robot to disable it. Several authors have been using a device called the Longpen to sign books remotely. Teleoperated robot aircraft, like the Predator Unmanned Aerial Vehicle, are increasingly being used by the military. These pilotless drones can search terrain and fire on targets. Hundreds of robots such as iRobot's Packbot and the Foster-Miller TALONare being used in Iraq and Afghanistan by the U.S. military to defuse roadside bombs or improvised explosive devices (IEDs) in an activity known as explosive ordnance disposal (EOD).

## Automated Fruit Harvesting Machines

Robots are used to automate picking fruit on orchards at a cost lower than that of human pickers.

## Domestic Robots

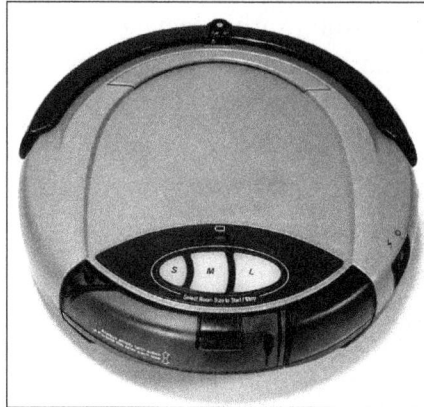

The Roomba domestic vacuum cleaner robot does a single, menial job.

Domestic robots are simple robots dedicated to a single task work in home use. They are used in simple but often disliked jobs, such as vacuum cleaning, floor washing, and lawn mowing. An example of a domestic robot is a Roomba.

## Military Robots

Military robots include the SWORDS robot which is currently used in ground-based combat. It can use a variety of weapons and there is some discussion of giving it some degree of autonomy in battleground situations.

Unmanned combat air vehicles (UCAVs), which are an upgraded form of UAVs, can do a wide variety of missions, including combat. UCAVs are being designed such as the BAE Systems Mantiswhich would have the ability to fly themselves, to pick their own course and target, and to make most decisions on their own. The BAE Taranis is a UCAV built by Great Britain which can fly across continents without a pilot and has new means to avoid detection. Flight trials are expected to begin in 2011.

The AAAI has studied this topic in depth and its president has commissioned a study to look at this issue.

Some have suggested a need to build "Friendly AI", meaning that the advances which are already occurring with AI should also include an effort to make AI intrinsically friendly and humane. Several such measures reportedly already exist, with robot-heavy countries such as Japan and South Korea having begun to pass regulations requiring robots to be equipped with safety systems, and possibly sets of 'laws' akin to Asimov's Three Laws of Robotics. An official report was issued in 2009 by the Japanese government's Robot Industry Policy Committee. Chinese officials and researchers have issued a report suggesting a set of ethical rules, and a set of new legal guidelines referred to as "Robot Legal Studies." Some concern has been expressed over a possible occurrence of robots telling apparent falsehoods.

## Mining Robots

Mining robots are designed to solve a number of problems currently facing the mining industry, including skills shortages, improving productivity from declining ore grades, and achieving environmental targets. Due to the hazardous nature of mining, in particular underground mining, the prevalence of autonomous, semi-autonomous, and tele-operated robots has greatly increased in recent times. A number of vehicle manufacturers provide autonomous trains, trucks and loaders that will load material, transport it on the mine site to its destination, and unload without requiring human intervention. One of the world's largest mining corporations, Rio Tinto, has recently expanded its autonomous truck fleet to the world's largest, consisting of 150 autonomous Komatsu trucks, operating in Western Australia. Similarly, BHP has announced the expansion of its autonomous drill fleet to the world's largest, 21 autonomous Atlas Copco drills.

Drilling, longwall and rockbreaking machines are now also available as autonomous robots. The Atlas Copco Rig Control System can autonomously execute a drilling plan on a drilling rig, moving the rig into position using GPS, set up the drill rig and drill down to specified depths. Similarly, the Transmin Rocklogic system can automatically plan a path to position a rockbreaker at a selected destination. These systems greatly enhance the safety and efficiency of mining operations.

## Healthcare

Robots in healthcare have two main functions. Those which assist an individual, such as a sufferer of a disease like Multiple Sclerosis, and those which aid in the overall systems such as pharmacies and hospitals.

## Home Automation for the Elderly and Disabled

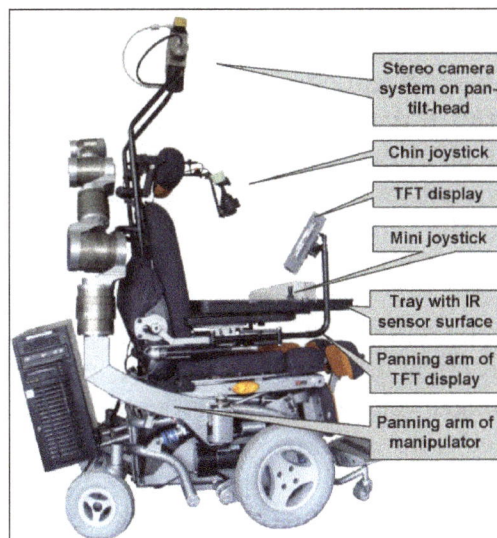

The Care-Providing Robot FRIEND.

Robots used in home automation have developed over time from simple basic robotic assistants, such as the Handy 1, through to semi-autonomous robots, such as FRIEND which can assist the elderly and disabled with common tasks.

The population is aging in many countries, especially Japan, meaning that there are increasing numbers of elderly people to care for, but relatively fewer young people to care for them. Humans make the best carers, but where they are unavailable, robots are gradually being introduced.

FRIEND is a semi-autonomous robot designed to support disabled and elderly people in their daily life activities, like preparing and serving a meal. FRIEND make it possible for patients who are paraplegic, have muscle diseases or serious paralysis (due to strokes etc.), to perform tasks without help from other people like therapists or nursing staff.

## Pharmacies

Script Pro manufactures a robot designed to help pharmacies fill prescriptions that consist of oral solids or medications in pill form. The pharmacist or pharmacy technician enters the prescription information into its information system. The system, upon determining whether or not the drug is in the robot, will send the information to the robot for filling. The robot has 3 different size vials to fill determined by the size of the pill. The robot technician, user, or pharmacist determines the needed size of the vial based on the tablet when the robot is stocked. Once the vial is filled it is brought up to a conveyor belt that delivers it to a holder that spins the vial and attaches the patient label. Afterwards it is set on another conveyor that delivers the patient's medication vial to a slot labeled with the patient's name on an LED read out. The pharmacist or technician then checks the contents of the vial to ensure it's the correct drug for the correct patient and then seals the vials and sends it out front to be picked up.

McKesson's Robot RX is another healthcare robotics product that helps pharmacies dispense thousands of medications daily with little or no errors. The robot can be ten feet wide and thirty feet long and can hold hundreds of different kinds of medications and thousands of doses. The pharmacy saves many resources like staff members that are otherwise unavailable in a resource scarce industry. It uses an electromechanical head coupled with a pneumatic system to capture each dose and deliver it to its either stocked or dispensed location. The head moves along a single axis while it rotates 180 degrees to pull the medications. During this process it uses barcode technology to verify its pulling the correct drug. It then delivers the drug to a patient specific bin on a conveyor belt. Once the bin is filled with all of the drugs that a particular patient needs and that the robot stocks, the bin is then released and returned out on the conveyor belt to a technician waiting to load it into a cart for delivery to the floor.

## Research Robots

While most robots today are installed in factories or homes, performing labour or life saving jobs, many new types of robot are being developed in laboratories around the world. Much of the research in robotics focuses not on specific industrial tasks, but on investigations into new types of robot, alternative ways to think about or design robots, and new ways to manufacture them. It is expected that these new types of robot will be able to solve real world problems when they are finally realized.

## Bionic and Biomimetic Robots

One approach to designing robots is to base them on animals. BionicKangaroo was designed and engineered by studying and applying the physiology and methods of locomotion of a kangaroo.

## Nanorobots

Nanorobotics is the emerging technology field of creating machines or robots whose components are at or close to the microscopic scale of a nanometer (10–9 meters). Also known as "nanobots" or "nanites", they would be constructed from molecular machines. So far, researchers have mostly produced only parts of these complex systems, such as bearings, sensors, and synthetic molecular motors, but functioning robots have also been made such as the entrants to the Nanobot Robocup contest. Researchers also hope to be able to create entire robots as small as viruses or bacteria, which could perform tasks on a tiny scale. Possible applications include micro surgery (on the level of individual cells), utility fog, manufacturing, weaponry and cleaning. Some people have suggested that if there were nanobots which could reproduce, the earth would turn into "grey goo", while others argue that this hypothetical outcome is nonsense.

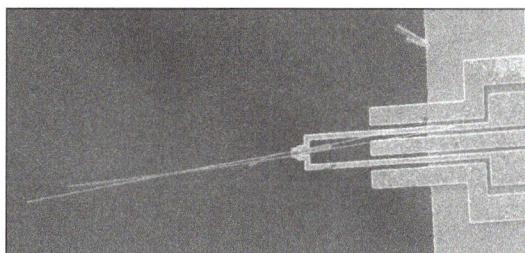

A microfabricated electrostatic gripper holding some silicon nanowires.

## Reconfigurable Robots

A few researchers have investigated the possibility of creating robots which can alter their physical form to suit a particular task, like the fictional T-1000. Real robots are nowhere near that sophisticated however, and mostly consist of a small number of cube shaped units, which can move relative to their neighbours. Algorithms have been designed in case any such robots become a reality.

## Soft-bodied Robots

Robots with silicone bodies and flexible actuators (air muscles, electroactive polymers, and ferrofluids) look and feel different from robots with rigid skeletons, and can have different behaviors. Soft, flexible (and sometimes even squishy) robots are often designed to mimic the biomechanics of animals and other things found in nature, which is leading to new applications in medicine, care giving, search and rescue, food handling and manufacturing, and scientific exploration.

## Swarm Robots

A swarm of robots from the open-source micro-robotic project.

Inspired by colonies of insects such as ants and bees, researchers are modeling the behavior of swarms of thousands of tiny robots which together perform a useful task, such as finding something hidden, cleaning, or spying. Each robot is quite simple, but the emergent behavior of the swarm is more complex. The whole set of robots can be considered as one single distributed system, in the same way an ant colony can be considered a superorganism, exhibiting swarm intelligence. The largest swarms so far created include the iRobot swarm, the SRI/MobileRobots CentiBots project and the Open-source Micro-robotic Project swarm, which are being used to research collective behaviors. Swarms are also more resistant to failure. Whereas one large robot may fail and ruin a mission, a swarm can continue even if several robots fail. This could make them attractive for space exploration missions, where failure is normally extremely costly.

## Haptic Interface Robots

Robotics also has application in the design of virtual reality interfaces. Specialized robots are in widespread use in the haptic research community. These robots, called "haptic interfaces", allow touch-enabled user interaction with real and virtual environments. Robotic forces allow simulating the mechanical properties of "virtual" objects, which users can experience through their sense of touch.

## Robots in Popular Culture

In 1942, science fiction writer Isaac Asimov's short story "Runaround" introduced the Three Laws of Robotics, which were said to be from the fictional "Handbook of Robotics" 56th edition, 2058. The three laws, at least according to some science fiction novels, are the only safety features required to ensure the safe operations of a robot:

1. A robot may not injure a human being or, through inaction, allow a human being to come to harm.

2. A robot must obey the orders given it by a human being except where such orders would conflict with the First Law.

3. A robot must protect its own existence as long as such protection does not conflict with the First or Second Laws.

## The Future

Numerous companies are working on consumer robots that can navigate their surroundings, recognize common objects, and perform simple chores without expert custom installation. Perhaps about the year 2020 the process will have produced the first broadly competent "universal robots" with lizardlike minds that can be programmed for almost any routine chore. With anticipated increases in computing power, by 2030 second-generation robots with trainable mouselike minds may become possible. Besides application programs, these robots may host a suite of software "conditioning modules" that generate positive- and negative-reinforcement signals in predefined circumstances.

By 2040 computing power should make third-generation robots with monkeylike minds possible. Such robots would learn from mental rehearsals in simulations that would model physical, cultural, and psychological factors. Physical properties would include shape, weight, strength, texture, and appearance of things and knowledge of how to handle them. Cultural aspects would include a thing's name, value, proper location, and purpose. Psychological factors, applied to humans and other robots, would include goals, beliefs, feelings, and preferences. The simulation would track external events and would tune its models to keep them faithful to reality. This should let a robot learn by imitation and afford it a kind of consciousness. By the middle of the 21st century, fourth-generation robots may exist with humanlike mental power able to abstract and generalize. Researchers hope that such machines will result from melding powerful reasoning programs to third-generation machines. Properly educated, fourth-generation robots are likely to become intellectually formidable.

# AEROBOT

The proposed Venus In-Situ Explorer lander would release a meteorology balloon.

An aerobot is an aerial robot, usually used in the context of an unmanned space probe or unmanned aerial vehicle.

While work has been done since the 1960s on robot "rovers" to explore the Moon and other worlds in the Solar system, such machines have limitations. They tend to be expensive and have limited range, and due to the communications time lags over interplanetary distances, they have to be smart enough to navigate without disabling themselves.

For planets with atmospheres of any substance, however, there is an alternative: an autonomous flying robot, or "aerobot". Most aerobot concepts are based on aerostats, primarily balloons, but occasionally airships. Flying above obstructions in the winds, a balloon could explore large regions of a planet in detail for relatively low cost. Airplanes for planetary exploration have also been proposed.

## Basics of Balloons

While the notion of sending a balloon to another planet sounds strange at first, balloons have a number of advantages for planetary exploration. They can be made light in weight and are potentially relatively inexpensive. They can cover a great deal of ground, and their view from a height gives them the ability to examine wide swathes of terrain with far more detail than would be available from an orbiting satellite. For exploratory missions, their relative lack of directional control is not a major obstacle as there is generally no need to direct them to a specific location.

Balloon designs for possible planetary missions have involved a few unusual concepts. One is the solar, or infrared (IR) Montgolfiere. This is a hot-air balloon where the envelope is made from a material that traps heat from sunlight, or from heat radiated from a planetary surface. Black is the best color for absorbing heat, but other factors are involved and the material may not necessarily be black.

Solar Montgolfieres have several advantages for planetary exploration, as they can be easier to deploy than a light gas balloon, do not necessarily require a tank of light gas for inflation, and are relatively forgiving of small leaks. They do have the disadvantage that they are only aloft during daylight hours.

The other is a "reversible fluid" balloon. This type of balloon consists of an envelope connected to a reservoir, with the reservoir containing a fluid that is easily vaporized. The balloon can be made to rise by vaporizing the fluid into gas, and can be made to sink by condensing the gas back into fluid. There are a number of different ways of implementing this scheme, but the physical principle is the same in all cases.

A balloon designed for planetary exploration will carry a small gondola containing an instrument payload. The gondola will also carry power, control, and communications subsystems. Due to weight and power supply constraints, the communications subsystem will generally be small and low power, and interplanetary communications will be performed through an orbiting planetary probe acting as a relay.

A solar Montgolfiere will sink at night, and will have a guide rope attached to the bottom of the gondola that will curl up on the ground and anchor the balloon during the darkness hours. The guide rope will be made of low friction materials to keep it from catching or tangling on ground features.

Alternatively, a balloon may carry a thicker instrumented "snake" in place of the gondola and guiderope, combining the functions of the two. This is a convenient scheme for making direct surface measurements.

A balloon could also be anchored to stay in one place to make atmospheric observations. Such a static balloon is known as an "aerostat".

One of the trickier aspects of planetary balloon operations is inserting them into operation. Typically, the balloon enters the planetary atmosphere in an "aeroshell", a heat shield in the shape of a flattened cone. After atmospheric entry, a parachute will extract the balloon assembly from the aeroshell, which falls away. The balloon assembly then deploys and inflates.

Once operational, the aerobot will be largely on its own and will have to conduct its mission autonomously, accepting only general commands over its long link to Earth. The aerobot will have to navigate in three dimensions, acquire and store science data, perform flight control by varying its altitude, and possibly make landings at specific sites to provide close-up investigation.

## The Venus Vega Balloons

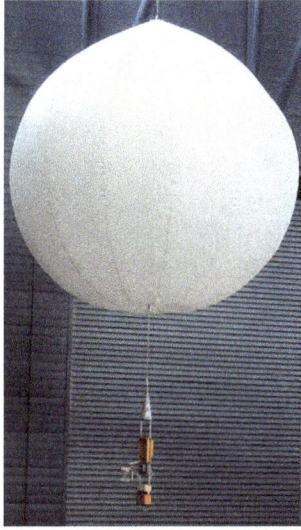

Vega balloon probe on display at the Udvar-Hazy Center of the Smithsonian Institution.

The first, and so far only, planetary balloon mission was performed by the Space Research Institute of Soviet Academy of Sciences in cooperation with the French space agency CNES in 1985. A small balloon, similar in appearance to terrestrial weather balloons, was carried on each of the two Soviet Vega Venus probes, launched in 1984.

The first balloon was inserted into the atmosphere of Venus on 11 June 1985, followed by the second balloon on 15 June 1985. The first balloon failed after only 56 minutes, but the second operated for a little under two Earth days until its batteries ran down.

The Venus Vega balloons were the idea of Jacques Blamont, chief scientist for CNES and the father of planetary balloon exploration. He energetically promoted the concept and enlisted international support for the small project.

The scientific results of the Venus VEGA probes were modest. More importantly, the clever and simple experiment demonstrated the validity of using balloons for planetary exploration.

## The Mars Aerobot Effort

After the success of the Venus VEGA balloons, Blamont focused on a more ambitious balloon mission to Mars, to be carried on a Soviet space probe.

The atmospheric pressure on Mars is about 150 times less than that of Earth. In such a thin atmosphere, a balloon with a volume of 5,000 to 10,000 cubic meters (178,500 to 357,000 cubic feet) could carry a payload of 20 kilograms (44 pounds), while a balloon with a volume of 100,000 cubic meters (3,600,000 cubic feet) could carry 200 kilograms (440 pounds).

The French had already conducted extensive experiments with solar Montgolfieres, performing over 30 flights from the late 1970s into the early 1990s. The Montgolfieres flew at an altitude of 35 kilometers, where the atmosphere was as thin and cold as it would be on Mars, and one spent 69 days aloft, circling the Earth twice.

Early concepts for the Mars balloon featured a "dual balloon" system, with a sealed hydrogen or helium-filled balloon tethered to a solar Montgolfiere. The light-gas balloon was designed to keep the Montgolfiere off the ground at night. During the day, the Sun would heat up the Montgolfiere, causing the balloon assembly to rise.

Eventually, the group decided on a cylindrical sealed helium balloon made of aluminized PET film, and with a volume of 5,500 cubic meters (196,000 cubic feet). The balloon would rise when heated during the day and sink as it cooled at night.

Total mass of the balloon assembly was 65 kilograms (143 pounds), with a 15 kilogram (33 pound) gondola and a 13.5 kilogram (30 pound) instrumented guiderope. The balloon was expected to operate for ten days. Unfortunately, although considerable development work was performed on the balloon and its subsystems, Russian financial difficulties pushed the Mars probe out from 1992, then to 1994, and then to 1996. The Mars balloon was dropped from the project due to cost.

## Planetary Aircraft

Artist's conception for a Venus airplane.

Winged airplane concepts have been proposed for robotic exploration in the atmosphere of Mars, Venus, Titan, and even Jupiter.

The main technical challenges of flying on Mars include:

1. Understanding and modeling the low Reynolds number, high subsonic Mach Number aerodynamics.

2. Building appropriate, often unconventional airframe designs and aerostructures.

3. Mastering the dynamics of deployment from a descending entry vehicle aeroshell.

4. Integrating a non-air-breathing propulsion subsystem into the system.

An aircraft concept, ARES was selected for a detailed design study as one of the four finalists for the 2007 Mars Scout Program opportunity, but was eventually not selected in favor of the Phoenix mission. In the design study, both half-scale and full-scale aircraft were tested under Mars-atmospheric conditions.

# AGRICULTURAL ROBOT

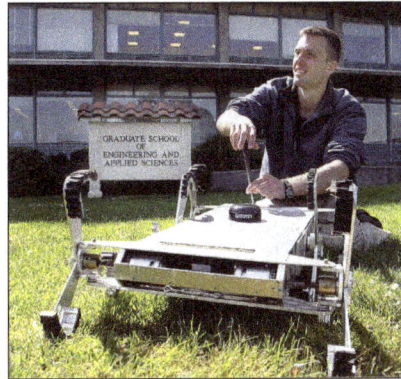
Autonomous Agricultural Robot.

An agricultural robot is a robot deployed for agricultural purposes. The main area of application of robots in agriculture today is at the harvesting stage. Emerging applications of robots or drones in agriculture include weed control, cloud seeding, planting seeds, harvesting, environmental monitoring and soil analysis. According to Verified Market Research, the agricultural robots market is expected to reach $11.58 billion by 2025.

## General

Fruit picking robots, driverless tractor / sprayers, and sheep shearing robots are designed to replace human labor. In most cases, a lot of factors have to be considered (e.g., the size and color of the fruit to be picked) before the commencement of a task. Robots can be used for other horticulturaltasks such as pruning, weeding, spraying and monitoring. Robots can also be used in livestockapplications (livestock robotics) such as automatic milking, washing and castrating. Robots like these have many benefits for the agricultural industry, including a higher quality of fresh produce, lower production costs, and a decreased need for manual labor. They can also be used to automate manual tasks, such as weed or bracken spraying, where the use of tractors and other manned vehicles is too dangerous for the operators.

## Designs

Fieldwork Robot.

The mechanical design consists of an end effector, manipulator, and gripper. Several factors must be considered in the design of the manipulator, including the task, economic efficiency, and required motions. The end effector influences the market value of the fruit and the gripper's design is based on the crop that is being harvested.

## End Effectors

An end effector in an agricultural robot is the device found at the end of the robotic arm, used for various agricultural operations. Several different kinds of end effectors have been developed. In an agricultural operation involving grapes in Japan, end effectors are used for harvesting, berry-thinning, spraying, and bagging. Each was designed according to the nature of the task and the shape and size of the target fruit. For instance, the end effectors used for harvesting were designed to grasp, cut, and push the bunches of grapes.

Berry thinning is another operation performed on the grapes, and is used to enhance the market value of the grapes, increase the grapes' size, and facilitate the bunching process. For berry thinning, an end effector consists of an upper, middle, and lower part. The upper part has two plates and a rubber that can open and close. The two plates compress the grapes to cut off the rachis branches and extract the bunch of grapes. The middle part contains a plate of needles, a compression spring, and another plate which has holes spread across its surface. When the two plates compress, the needles punch holes through the grapes. Next, the lower part has a cutting device which can cut the bunch to standardize its length.

For spraying, the end effector consists of a spray nozzle that is attached to a manipulator. In practice, producers want to ensure that the chemical liquid is evenly distributed across the bunch. Thus, the design allows for an even distribution of the chemical by making the nozzle to move at a constant speed while keeping distance from the target.

The final step in grape production is the bagging process. The bagging end effector is designed with a bag feeder and two mechanical fingers. In the bagging process, the bag feeder is composed of slits which continuously supply bags to the fingers in an up and down motion. While the bag is being fed to the fingers, two leaf springs that are located on the upper end of the bag hold the bag open. The bags are produced to contain the grapes in bunches. Once the bagging process is complete, the fingers open and release the bag. This shuts the leaf springs, which seals the bag and prevents it from opening again.

## Gripper

The gripper is a grasping device that is used for harvesting the target crop. Design of the gripper is based on simplicity, low cost, and effectiveness. Thus, the design usually consists of two mechanical fingers that are able to move in synchrony when performing their task. Specifics of the design depend on the task that is being performed. For example, in a procedure that required plants to be cut for harvesting, the gripper was equipped with a sharp blade.

## Manipulator

The manipulator allows the gripper and end effector to navigate through their environment. The manipulator consists of four-bar parallel links that maintain the gripper's position and height. The manipulator also can utilize one, two, or three pneumatic actuators. Pneumatic actuatorsare motors which produce linear and rotary motion by converting compressed air into energy. The pneumatic actuator is the most effective actuator for agricultural robots because of its high power-weight ratio. The most cost efficient design for the manipulator is the single actuator configuration, yet this is the least flexible option.

## Development

The first development of robotics in agriculture can be dated as early as the 1920s, with research to incorporate automatic vehicle guidance into agriculture beginning to take shape. This research led to the advancements between the 1950s and 60s of autonomous agricultural vehicles. The concept was not perfect however, with the vehicles still needing a cable system to guide their path. Robots in agriculture continued to develop as technologies in other sectors began to develop as well. It was not until the 1980s, following the development of the computer, that machine vision guidance became possible.

Other developments over the years included the harvesting of oranges using a robot both in France and the US.

While robots have been incorporated in indoor industrial settings for decades, outdoor robots for the use of agriculture are considered more complex and difficult to develop. This is due to concerns over safety, but also over the complexity of picking crops subject to different environmental factors and unpredictability.

## Demand in the Market

There are concerns over the amount of labor the agricultural sector needs. With an aging population, Japan is unable to meet the demands of the agricultural labor market. Similarly, the United State currently depends on a large number of immigrant workers, but between the decrease in seasonal farmworkers and increased efforts to stop immigration by the government, they too are unable to meet the demand. Businesses are often forced to let crops rot due to an inability to pick them all by the end of the season. Additionally, there are concerns over the growing population that will need to be fed over the next years. Because of this, there is a large desire to improve agricultural machinery to make it more cost efficient and viable for continued use.

## Current Applications and Trends

Much of the current research continues to work towards autonomous agricultural vehicles. This research is based on the advancements made in driver-assist systems and self-driving cars.

While robots have already been incorporated in many areas of agricultural farm work, they are still largely missing in the harvest of various crops. This has started to change as companies begin to develop robots that complete more specific tasks on the farm. The biggest concern over robots harvesting crops comes from harvesting soft crops such as strawberries which can easily be damaged or missed entirely. Despite these concerns, progress in this area is being made. According to Gary Wishnatzki, the co-founder of Harvest Croo Robotics, one of their strawberry pickers currently being tested in Florida can "pick a 25-acre field in just three days and replace a crew of about 30 farm workers". Similar progress is being made in harvesting apples, grapes, and other crops.

Another goal being set by agricultural companies involves the collection of data. There are rising concerns over the growing population and the decreasing labor available to feed them. Data collection is being developed as a way to increase productivity on farms. AgriData is currently developing new technology to do just this and help farmers better determine the best time to harvest their crops by scanning fruit trees.

## Applications

Robots have many fields of application in agriculture. Some examples and prototypes of robots include the Merlin Robot Milker, Rosphere, Harvest Automation, Orange Harvester, lettuce bot, and weeder. One case of a large scale use of robots in farming is the milk bot. It is widespread among British dairy farms because of its efficiency and nonrequirement to move. According to David Gardner (chief executive of the Royal Agricultural Society of England), a robot can complete a complicated task if its repetitive and the robot is allowed to sit in a single place. Furthermore, robots that work on repetitive tasks (e.g. milking) fulfill their role to a consistent and particular standard.

Another field of application is horticulture. One horticultural application is the development of RV100 by Harvest Automation Inc. RV 100 is designed to transport potted plants in a greenhouse or outdoor setting. The functions of RV100 in handling and organizing potted plants include spacing capabilities, collection, and consolidation. The benefits of using RV100 for this task include high placement accuracy, autonomous outdoor and indoor function, and reduced production costs.

Examples:

- Vinobot and Vinoculer.

- LSU's AgBot.

- Harvest Automation is a company founded by former iRobot employees to develop robots for greenhouses.

- Root AI has made a tomato-picking robot for use in greenhouses.

- Strawberry picking robot from Robotic Harvesting and Agrobot.

- Small Robot Company developed a range of small agricultural robots, each one being focused on a particular task (weeding, spraying, drilling holes) and controlled by an AI system.

- EcoRobotix has made a solar-powered weeding and spraying robot.

- Blue River Technology has developed a farm implement for a tractor which only sprays plants that require spraying, reducing herbicide use by 90%.

- Casmobot next generation slope mower.

- Fieldrobot Event is a competition in mobile agricultural robotics.

- HortiBot - A Plant Nursing Robot.

- Lettuce Bot - Organic Weed Elimination and Thinning of Lettuce.

- Rice planting robot developed by the Japanese National Agricultural Research Centre.

- ROS Agriculture - Open source software for agricultural robots using the Robot Operating System.

- The IBEX autonomous weed spraying robot for extreme terrain, under development.

- FarmBot, Open Source CNC Farming.

- VAE, under development by an argentinean ag-tech startup, aims to become a universal platform for multiple agricultural applications, from precision spraying to livestock handling.

- ACFR RIPPA: for spot spraying.

- ACFR SwagBot; for livestock monitoring.

- ACFR Digital Farmhand: for spraying, weeding and seeding.

# AUTONOMOUS ROBOT

An autonomous robot is a robot that performs behaviors or tasks with a high degree of autonomy (without external influence). Autonomous robotics is usually considered to be a subfield of artificial intelligence, robotics, and information engineering. Early versions were proposed and demonstrated by author/inventor David L. Heiserman.

Autonomous robots are particularly desirable in fields such as spaceflight, household maintenance (such as cleaning), waste water treatment, and delivering goods and services.

Some modern factory robots are "autonomous" within the strict confines of their direct environment. It may not be that every degree of freedom exists in their surrounding environment, but the factory robot's workplace is challenging and can often contain chaotic, unpredicted variables. The exact orientation and position of the next object of work and (in the more advanced factories) even the type of object and the required task must be determined. This can vary unpredictably (at least from the robot's point of view).

One important area of robotics research is to enable the robot to cope with its environment whether this be on land, underwater, in the air, underground, or in space.

A fully autonomous robot can:

- Gain information about the environment.

- Work for an extended period without human intervention.

- Move either all or part of itself throughout its operating environment without human assistance.

- Avoid situations that are harmful to people, property, or itself unless those are part of its design specifications.

An autonomous robot may also learn or gain new knowledge like adjusting for new methods of accomplishing its tasks or adapting to changing surroundings.

Like other machines, autonomous robots still require regular maintenance.

## Components and Criteria of Robotic Autonomy

### Self-maintenance

The first requirement for complete physical autonomy is the ability for a robot to take care of itself. Many of the battery-powered robots on the market today can find and connect to a charging station, and some toys like Sony's *Aibo* are capable of self-docking to charge their batteries.

Self-maintenance is based on "proprioception", or sensing one's own internal status. In the battery charging example, the robot can tell proprioceptively that its batteries are low and it then seeks the charger. Another common proprioceptive sensor is for heat monitoring. Increased proprioception will be required for robots to work autonomously near people and in harsh environments. Common proprioceptive sensors include thermal, optical, and haptic sensing, as well as the Hall effect (electric).

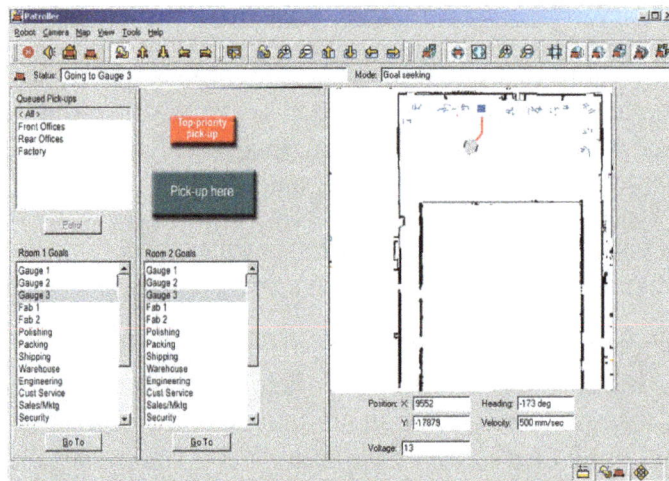

Robot GUI display showing battery voltage and other proprioceptive data in lower right-hand corner. The display is for user information only. Autonomous robots monitor and respond to proprioceptive sensors without human intervention to keep themselves safe and operating properly.

### Sensing the Environment

Exteroception is sensing things about the environment. Autonomous robots must have a range of environmental sensors to perform their task and stay out of trouble.

- Common exteroceptive sensors include the electromagnetic spectrum, sound, touch, chemical (smell, odor), temperature, range to various objects, and altitude.

Some robotic lawn mowers will adapt their programming by detecting the speed in which grass grows as needed to maintain a perfectly cut lawn, and some vacuum cleaning robots have dirt detectors that sense how much dirt is being picked up and use this information to tell them to stay in one area longer.

## Task Performance

The next step in autonomous behavior is to actually perform a physical task. A new area showing commercial promise is domestic robots, with a flood of small vacuuming robots beginning with iRobot and Electrolux in 2002. While the level of intelligence is not high in these systems, they navigate over wide areas and pilot in tight situations around homes using contact and non-contact sensors. Both of these robots use proprietary algorithms to increase coverage over simple random bounce.

The next level of autonomous task performance requires a robot to perform conditional tasks. For instance, security robots can be programmed to detect intruders and respond in a particular way depending upon where the intruder is.

## Autonomous Navigation

### Indoor Navigation

For a robot to associate behaviors with a place (localization) requires it to know where it is and to be able to navigate point-to-point. Such navigation began with wire-guidance in the 1970s and progressed in the early 2000s to beacon-based triangulation. Current commercial robots autonomously navigate based on sensing natural features. The first commercial robots to achieve this were Pyxus' HelpMate hospital robot and the CyberMotion guard robot, both designed by robotics pioneers in the 1980s. These robots originally used manually created CAD floor plans, sonar sensing and wall-following variations to navigate buildings. The next generation, such as MobileRobots' PatrolBotand autonomous wheelchair, both introduced in 2004, have the ability to create their own laser-based maps of a building and to navigate open areas as well as corridors. Their control system changes its path on the fly if something blocks the way.

At first, autonomous navigation was based on planar sensors, such as laser range-finders, that can only sense at one level. The most advanced systems now fuse information from various sensors for both localization (position) and navigation. Systems such as Motivity can rely on different sensors in different areas, depending upon which provides the most reliable data at the time, and can re-map a building autonomously.

Rather than climb stairs, which requires highly specialized hardware, most indoor robots navigate handicapped-accessible areas, controlling elevators, and electronic doors. With such electronic access-control interfaces, robots can now freely navigate indoors. Autonomously climbing stairs and opening doors manually are topics of research at the current time.

As these indoor techniques continue to develop, vacuuming robots will gain the ability to clean a specific user-specified room or a whole floor. Security robots will be able to cooperatively surround intruders and cut off exits. These advances also bring concomitant protections: robots' internal maps typically permit "forbidden areas" to be defined to prevent robots from autonomously entering certain regions.

### Outdoor Navigation

Outdoor autonomy is most easily achieved in the air, since obstacles are rare. Cruise missiles are rather dangerous highly autonomous robots. Pilotless drone aircraft are increasingly used for reconnaissance. Some of these unmanned aerial vehicles (UAVs) are capable of flying their entire mission without any human interaction at all except possibly for the landing where a person intervenes using radio remote control. Some drones are capable of safe, automatic landings, however. An autonomous ship was announced in 2014—the Autonomous spaceport drone ship—and is scheduled to make its first operational test in December 2014.

Outdoor autonomy is the most difficult for ground vehicles, due to:

- Three-dimensional terrain.

- Great disparities in surface density.

- Weather exigencies.

- Instability of the sensed environment.

## Military Robot

Armed Predator drone.

Military robots are autonomous robots or remote-controlled mobile robots designed for military applications, from transport to search & rescue and attack.

Some such systems are currently in use, and many are under development.

Examples:

## In Current Use

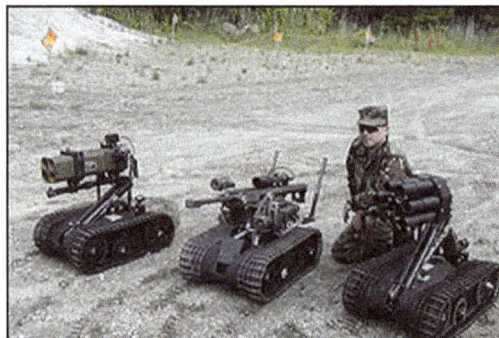

Foster-Miller TALON SWORDS units equipped with various weaponry.

The Platforma-M variant of the Multifunctional Utility/Combat
support/Patrol. Serially produced by the Russian Army.

- DRDO Daksh

- D9T Panda, Israel

- Elbit Hermes 450, Israel

- Goalkeeper CIWS

- Guardium

- IAIO Fotros, Iran

- PackBot

- MQ-9 Predator B

- MQ-1 Predator

- TALON

- Samsung SGR-A1

- Shahed 129, Iran

- Shomer Gvouloth ( "Border Keeper"), Israel

## In Development

The Armed Robotic Vehicle variant of the MULE. Image made by the U.S. Army.

- US Mechatronics has produced a working automated sentry gun and is currently developing it further for commercial and military use.

- MIDARS, a four-wheeled robot outfitted with several cameras, radar, and possibly a firearm, that automatically performs random or preprogrammed patrols around a military base or other government installation. It alerts a human overseer when it detects movement in unauthorized areas, or other programmed conditions. The operator can then instruct the robot to ignore the event, or take over remote control to deal with an intruder, or to get better camera views of an emergency. The robot would also regularly scan radio frequency identification tags (RFID) placed on stored inventory as it passed and report any missing items.

- Tactical Autonomous Combatant (TAC) units, described in Project Alpha study *Unmanned Effects: Taking the Human out of the Loop*.

- Autonomous Rotorcraft Sniper System is an experimental robotic weapons system being developed by the U.S. Army since 2005. It consists of a remotely operated sniper rifleattached to an unmanned autonomous helicopter. It is intended for use in urban combat or for several other missions requiring snipers. Flight tests are scheduled to begin in summer 2009.

- The "Mobile Autonomous Robot Software" research program was started in December 2003 by the Pentagon who purchased 15 Segways in an attempt to develop more advanced military robots. The program was part of a $26 million Pentagon program to develop software for autonomous systems.

- ACER.

- Atlas (robot).

- Battlefield Extraction-Assist Robot.

- Dassault nEUROn (French UCAV).

- Dragon Runner.

- MATILDA.

- MULE (US UGV).

- R-Gator.

- Ripsaw MS1.

- SUGV.

- Syrano.

- iRobot Warrior.

- PETMAN.

- Excalibur unmanned aerial vehicle.

## Effects and Impact

### Advantages

Autonomous robotics would save and preserve soldiers' lives by removing serving soldiers, who might otherwise be killed, from the battlefield. Lt. Gen. Richard Lynch of the United States Army Installation Management Command and assistant Army chief of staff for installation stated at a conference:

As we think about what's happening on the battlefield today. We contend there are things we could do to improve the survivability of our service members. And we all know that's true.

Major Kenneth Rose of the US Army's Training and Doctrine Command outlined some of the advantages of robotic technology in warfare:

Machines don't get tired. They don't close their eyes. They don't hide under trees when it rains and they don't talk to their friends. A human's attention to detail on guard duty drops dramatically in the first 30 minutes. Machines know no fear.

Increasing attention is also paid to how to make the robots more autonomous, with a view of eventually allowing them to operate on their own for extended periods of time, possibly behind enemy lines. For such functions, systems like the Energetically Autonomous Tactical Robot are being tried, which is intended to gain its own energy by foraging for plant matter. The majority of military robots are tele-operated and not equipped with weapons; they are used for reconnaissance, surveillance, sniper detection, neutralizing explosive devices, etc. Current robots that are equipped with weapons are tele-operated so they are not capable of taking lives autonomously. Advantages regarding the lack of emotion and passion in robotic combat is also taken into consideration as a beneficial factor in significantly reducing instances of unethical behavior in wartime. Autonomous machines are created not to be "truly 'ethical' robots", yet ones that comply with the laws of war (LOW) and rules of engagement (ROE). Hence the fatigue, stress, emotion, adrenaline, etc. that affect a human soldier's rash decisions are removed; there will be no effect on the battlefield caused by the decisions made by the individual.

### Risks

Human rights groups and NGOs such as Human Rights Watch and the Campaign to Stop Killer Robots have started urging governments and the United Nations to issue policy to outlaw the development of so-called "lethal autonomous weapons systems" (LAWS). The United Kingdom opposed such campaigns, with the Foreign Office declaring that "international humanitarian law already provides sufficient regulation for this area".

In July 2015, over 1,000 experts in artificial intelligence signed a letter calling for a ban on autonomous weapons. The letter was presented in Buenos Aires at the 24th International Joint Conference on Artificial Intelligence (IJCAI-15) and was co-signed by Stephen Hawking, Elon Musk, Steve Wozniak, Noam Chomsky, Skype co-founder Jaan Tallinn and Google DeepMind co-founder Demis Hassabis, among others.

### Psychology

American soldiers have been known to name the robots that serve alongside them. These names are often in honor of human friends, family, celebrities, pets, or are eponymic. The 'gender' assigned to the robot may be related to the marital status of its operator.

Some affixed fictitious medals to battle-hardened robots, and even held funerals for destroyed robots. An interview of 23 explosive ordnance detection members shows that while they feel it is better to lose a robot than a human, they also felt anger and a sense of loss if they were destroyed. A survey of 746 people in the military showed that 80% either 'liked' or 'loved' their military robots, with more affection being shown towards ground rather than aerial robots. Surviving dangerous combat situations together increased the level of bonding between soldier and robot, and current and future advances in artificial intelligence may further intensify the bond with the military robots.

# MOBILE ROBOT

A mobile robot is a robot that is capable of locomotion. Mobile robotics is usually considered to be a subfield of robotics and information engineering.

A spying robot is an example of a mobile robot capable of movement in a given environment.

Mobile robots have the capability to move around in their environment and are not fixed to one physical location. Mobile robots can be "autonomous" (AMR - autonomous mobile robot) which means they are capable of navigating an uncontrolled environment without the need for physical or electro-mechanical guidance devices. Alternatively, mobile robots can rely on guidance devices that allow them to travel a pre-defined navigation route in relatively controlled space (AGV - autonomous guided vehicle). By contrast, industrial robots are usually more-or-less stationary, consisting of a jointed arm (multi-linked manipulator) and gripper assembly (or end effector), attached to a fixed surface.

Mobile robots have become more commonplace in commercial and industrial settings. Hospitals have been using autonomous mobile robots to move materials for many years. Warehouses have installed mobile robotic systems to efficiently move materials from stocking shelves to order fulfillment zones. Mobile robots are also a major focus of current research and almost every major university has one or more labs that focus on mobile robot research. Mobile robots are also found in industrial, military and security settings. Domestic robots are consumer products, including entertainment robots and those that perform certain household tasks such as vacuuming or gardening.

The components of a mobile robot are a controller, control software, sensors and actuators. The controller is generally a microprocessor, embedded microcontroller or a personal computer (PC). Mobile control software can be either assembly level language or high-level languages such as C, C++, Pascal, Fortran or special real-time software. The sensors used are dependent upon the requirements of the robot. The requirements could be dead reckoning, tactile and proximity sensing, triangulation ranging, collision avoidance, position location and other specific applications.

## Classification

Mobile robots may be classified by:

- The environment in which they travel:

  ◦ Land or home robots are usually referred to as Unmanned Ground Vehicles (UGVs). They are most commonly wheeled or tracked, but also include legged robots with two or more legs (humanoid, or resembling animals or insects).

  ◦ Delivery & Transportation robots can move materials and supplies through a work environment.

  ◦ Aerial robots are usually referred to as Unmanned Aerial Vehicles (UAVs).

  ◦ Underwater robots are usually called autonomous underwater vehicles (AUVs).

  ◦ Polar robots, designed to navigate icy, crevasse filled environments.

- The device they use to move, mainly:

  ◦ Legged robot : human-like legs (i.e. an android) or animal-like legs.

  ◦ Wheeled robot.

  ◦ Tracks.

## Mobile Robot Navigation

There are many types of mobile robot navigation:

### Manual Remote or Tele-op

A manually teleoperated robot is totally under control of a driver with a joystick or other control device. The device may be plugged directly into the robot, may be a wireless joystick, or may be an accessory to a wireless computer or other controller. A tele-op'd robot is typically used to keep the operator out of harm's way. Examples of manual remote robots include Robotics Design's ANATROLLER ARI-100 and ARI-50, Foster-Miller's Talon, iRobot's PackBot, and KumoTek's MK-705 Roosterbot.

### Guarded Tele-op

A guarded tele-op robot has the ability to sense and avoid obstacles but will otherwise navigate as driven, like a robot under manual tele-op. Few if any mobile robots offer only guarded tele-op.

## Line-following Car

Some of the earliest Automated Guided Vehicles (AGVs) were line following mobile robots. They might follow a visual line painted or embedded in the floor or ceiling or an electrical wire in the floor. Most of these robots operated a simple "keep the line in the center sensor" algorithm. They could not circumnavigate obstacles; they just stopped and waited when something blocked their path. Many examples of such vehicles are still sold, by Transbotics, FMC, Egemin, HK Systems and many other companies. These types of robots are still widely popular in well known Robotic societies as a first step towards learning nooks and corners of robotics.

## Autonomously Randomized Robot

Autonomous robots with random motion basically bounce off walls, whether those walls are sensed.

## Autonomously Guided Robot

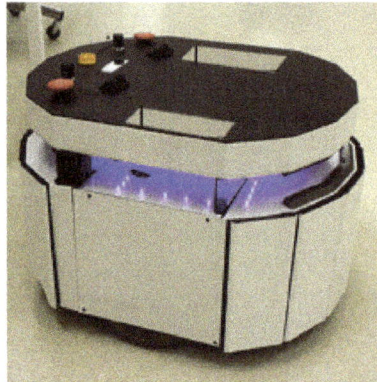

Robot developers use ready-made autonomous bases and software to design robot applications quickly.
Shells shaped like people or cartoon characters may cover the base to disguise it.

An autonomously guided robot knows at least some information about where it is and how to reach various goals and or waypoints along the way. "Localization" or knowledge of its current location, is calculated by one or more means, using sensors such motor encoders, vision, Stereopsis, lasers and global positioning systems. Positioning systems often use triangulation, relative position and/or Monte-Carlo/Markov localization to determine the location and orientation of the platform, from which it can plan a path to its next waypoint or goal. It can gather sensor readings that are time- and location-stamped. Such robots are often part of the wireless enterprise network, interfaced with other sensing and control systems in the building. For instance, the PatrolBot security robot responds to alarms, operates elevators and notifies the command center when an incident arises. Other autonomously guided robots include the SpeciMinder and the TUG delivery robots for the hospital. In 2013, autonomous robots capable of finding sunlight and water for potted plants were created by artist Elizabeth Demaray in collaboration with engineer Dr. Qingze Zou, biologist Dr. Simeon Kotchomi, and computer scientist Dr. Ahmed Elgammal.

## Sliding Autonomy

More capable robots combine multiple levels of navigation under a system called sliding autonomy.

Most autonomously guided robots, such as the HelpMate hospital robot, also offer a manual mode. The Motivity autonomous robot operating system, which is used in the ADAM, PatrolBot, SpeciMinder, MapperBot and a number of other robots, offers full sliding autonomy, from manual to guarded to autonomous modes.

# LEGGED ROBOT

A hexapod robot.

Legged robots are a type of mobile robot, which use articulated limbs, such as leg mechanisms, to provide locomotion. They are more versatile than wheeled robots and can traverse many different terrains, though these advantages require increased complexity and power consumption. Legged robots often imitate legged animals, such as humans or insects, in an example of biomimicry.

## Gait and Support Pattern

Legged robots, or walking machines, are designed for locomotion on rough terrain and require control of leg actuators to maintain balance, sensors to determine foot placement and planning algorithms to determine the direction and speed of movement. The periodic contact of the legs of the robot with the ground is called the gait of the walker.

In order to maintain locomotion the center of gravity of the walker must be supported either statically or dynamically. Static support is provided by ensuring the center of gravity is within the support pattern formed by legs in contact with the ground. Dynamic support is provided by keeping the trajectory of the center of gravity located so that it can be repositioned by forces from one or more of its legs.

## Types

Legged robots can be categorized by the number of limbs they use, which determines gaits available. Many-legged robots tend to be more stable, while fewer legs lends itself to greater maneuverability.

## One-legged

One-legged, or pogo stick robots use a hopping motion for navigation. In the 1980s, Carnegie Mellon University developed a one-legged robot to study balance. Berkeley's SALTO is another example.

## Two-legged

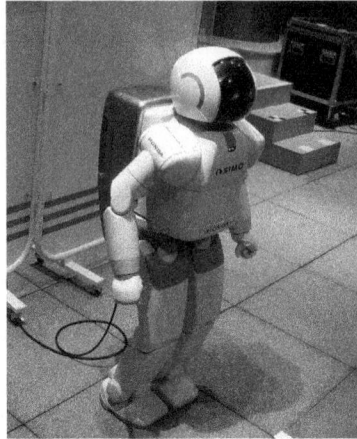

ASIMO - a bipedal robot.

*Bipedal* or two-legged robots exhibit bipedal motion. As such, they face two primary problems:

- Stability control, which refers to a robot's balance.

- Motion control, which refers to a robot's ability to move.

Stability control is particularly difficult for bipedal systems, which must maintain balance in the forward-backward direction even at rest. Some robots, especially toys, solve this problem with large feet, which provide greater stability while reducing mobility. Alternatively, more advanced systems use sensors such as accelerometers or gyroscopes to provide dynamic feedback in a fashion that approximates a human being's balance. Such sensors are also employed for motion control and walking. The complexity of these tasks lends itself to machine learning.

Simple bipedal motion can be approximated by a rolling polygon where the length of each side matches that of a single step. As the step length grows shorter, the number of sides increases and the motion approaches that of a circle. This connects bipedal motion to wheeled motion as a limit of stride length.

Two-legged robots include:

- Boston Dynamics' Atlas.

- Toy robots such as QRIO and ASIMO.

- NASA's Valkyrie robot, intended to aid humans on Mars.

- The ping-pong playing TOPIO robot.

## Four-legged

*Quadrupedal* or four-legged robots exhibit quadrupedal motion. They benefit from increased stability over bipedal robots, especially during movement. At slow speeds, a quadrupedal robot may move only one leg at a time, ensuring a stable tripod. Four-legged robots also benefit from a lower center of gravity than two-legged systems.

Quadruped robot "Big Dog" is being developed as a mule that can traverse difficult terrain.

Four legged robots include:

- The TITAN series, developed since the 1980s by the Hirose-Yoneda Laboratory.

- The dynamically stable BigDog, developed in 2005 by Boston Dynamics, NASA's Jet Propulsion Laboratory, and the Harvard University Concord Field Station.

- BigDog's successor, the LS3.

## Six-legged

Six-legged robots, or *hexapods*, are motivated by a desire for even greater stability than bipedal or quadrupedal robots. Their final designs often mimic the mechanics of insects, and their gaits may be categorized similarly. These include:

- Wave gait: the slowest gait, in which pairs of legs move in a "wave" from the back to the front.

- Tripod gait: a slightly faster step, in which three legs move at once. The remaining three legs provide a stable tripod for the robot.

Six-legged robots include:

- Odex, a 375-pound hexapod developed by Odetics in the 1980s. Odex distinguished itself with its onboard computers, which controlled each leg.

- Genghis, one of the earliest autonomous six-legged robots, was developed at MIT by Rodney Brooks in the 1980s.

- The modern toy series, Hexbug.

## Eight-legged

Eight-legged legged robots are inspired by spiders and other arachnids, as well as some underwater walkers. They offer by far the greatest stability, which enabled some early successes with legged robots.

Eight-legged robots include:

- Dante, a Carnegie Mellon University project designed to explore Mount Erebus.

- The T8X, a commercially available robot designed to emulate a spider's appearance and movements.

## Hybrids

Some robots use a combination of legs and wheels. This grants a machine the speed and energy efficiency of wheeled locomotion as well as the mobility of legged navigation. Boston Dynamics' Handle, a bipedal robot with wheels on both legs, is one example.

# HEXAPOD ROBOT

Beetle hexapod.

A hexapod robot is a mechanical vehicle that walks on six legs. Since a robot can be statically stable on three or more legs, a hexapod robot has a great deal of flexibility in how it can move. If legs become disabled, the robot may still be able to walk. Furthermore, not all of the robot's legs are needed for stability; other legs are free to reach new foot placements or manipulate a payload.

Many hexapod robots are biologically inspired by Hexapoda locomotion. Hexapods may be used to test biological theories about insect locomotion, motor control, and neurobiology.

## Designs

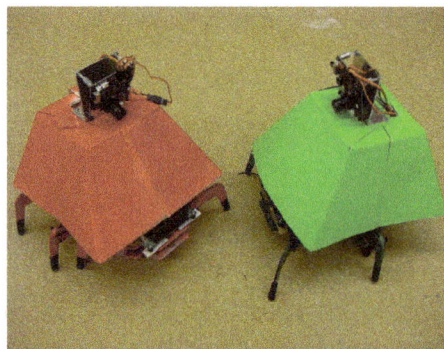

Two hexapod robots at the Georgia Institute of Technology with CMUCams mounted on top.

Hexapod designs vary in leg arrangement. Insect-inspired robots are typically laterally symmetric, such as the RiSE robot at Carnegie Mellon. A radially symmetric hexapod is ATHLETE (All-Terrain Hex-Legged Extra-Terrestrial Explorer) robot at JPL.

Typically, individual legs range from two to six degrees of freedom. Hexapod feet are typically pointed, but can also be tipped with adhesive material to help climb walls or wheels so the robot can drive quickly when the ground is flat.

## Locomotion

Walking hexapod simulated in Webots.

Most often, hexapods are controlled by gaits, which allow the robot to move forward, turn, and perhaps side-step. Some of the most common gaits are as follows:

- Alternating tripod: 3 legs on the ground at a time.

- Quadruped.

- Crawl: move just one leg at a time.

Gaits for hexapods are often stable, even in slightly rocky and uneven terrain.

Motion may also be nongaited, which means the sequence of leg motions is not fixed, but rather chosen by the computer in response to the sensed environment. This may be most helpful in very rocky terrain, but existing techniques for motion planning are computationally expensive.

## Biologically Inspired

Insects are chosen as models because their nervous system are simpler than other animal species. Also, complex behaviours can be attributed to just a few neurons and the pathway between sensory input and motor output is relatively shorter. Insects' walking behaviour and neural architecture are used to improve robot locomotion. Conversely, biologists can use hexapod robots for testing different hypotheses.

Biologically inspired hexapod robots largely depend on the insect species used as a model. The cockroach and the stick insect are the two most commonly used insect species; both have been ethologically and neurophysiologically extensively studied. At present no complete nervous system is known, therefore, models usually combine different insect models, including those of other insects.

Insect gaits are usually obtained by two approaches: the centralized and the decentralized control architectures. Centralized controllers directly specify transitions of all legs, whereas in decentralized architectures, six nodes (legs) are connected in a parallel network; gaits arise by the interaction between neighbouring legs.

# HUMANOID ROBOT

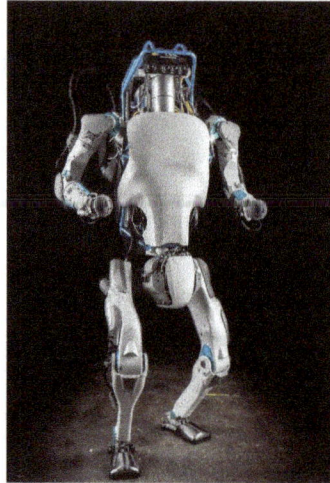

Atlas from Boston Dynamics.

A humanoid robot is a robot with its body shape built to resemble the human body. The design may be for functional purposes, such as interacting with human tools and environments, for experimental purposes, such as the study of bipedal locomotion, or for other purposes. In general, humanoid robots have a torso, a head, two arms, and two legs, though some forms of humanoid robots may model only part of the body, for example, from the waist up. Some humanoid robots also have heads designed to replicate human facial features such as eyes and mouths. Androids are humanoid robots built to aesthetically resemble humans.

## Purpose

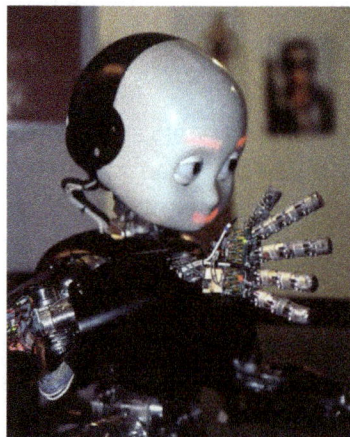

iCub robot at the Genoa Science Festival, Italy, in 2009.

Valkyrie, from NASA.

Humanoid robots are now used as research tools in several scientific areas. Researchers study the human body structure and behavior (biomechanics) to build humanoid robots. On the other side, the attempt to simulate the human body leads to a better understanding of it. Human cognition is a field of study which is focused on how humans learn from sensory information in order to acquire perceptual and motor skills. This knowledge is used to develop computational models of human behavior and it has been improving over time.

It has been suggested that very advanced robotics will facilitate the enhancement of ordinary humans.

Although the initial aim of humanoid research was to build better orthosis and prosthesis for human beings, knowledge has been transferred between both disciplines. A few examples are powered leg prosthesis for neuromuscularly impaired, ankle-foot orthosis, biological realistic leg prosthesis and forearm prosthesis.

Besides the research, humanoid robots are being developed to perform human tasks like personal assistance, through which they should be able to assist the sick and elderly, and dirty or dangerous jobs. Humanoids are also suitable for some procedurally-based vocations, such as reception-desk administrators and automotive manufacturing line workers. In essence, since they can use tools and operate equipment and vehicles designed for the human form, humanoids could theoretically perform any task a human being can, so long as they have the proper software. However, the complexity of doing so is immense.

They are also becoming increasingly popular as entertainers. For example, Ursula, a female robot, sings, plays music, dances and speaks to her audiences at Universal Studios. Several Disney theme park shows utilize animatronic robots that look, move and speak much like human beings. Although these robots look realistic, they have no cognition or physical autonomy. Various humanoid robots and their possible applications in daily life are featured in an independent documentary film called *Plug & Pray*, which was released in 2010.

Humanoid robots, especially those with artificial intelligence algorithms, could be useful for future dangerous and/or distant space exploration missions, without having the need to turn back around again and return to Earth once the mission is completed.

## Sensors

A sensor is a device that measures some attribute of the world. Being one of the three primitives of robotics (besides planning and control), sensing plays an important role in robotic paradigms.

Sensors can be classified according to the physical process with which they work or according to the type of measurement information that they give as output. In this case, the second approach was used.

## Proprioceptive Sensors

Proprioceptive sensors sense the position, the orientation and the speed of the humanoid's body and joints.

In human beings the otoliths and semi-circular canals (in the inner ear) are used to maintain balance and orientation. In addition humans use their own proprioceptive sensors (e.g. touch, muscle extension, limb position) to help with their orientation. Humanoid robots use accelerometers to measure the acceleration, from which velocity can be calculated by integration; tilt sensors to measure inclination; force sensors placed in robot's hands and feet to measure contact force with environment; position sensors, that indicate the actual position of the robot (from which the velocity can be calculated by derivation) or even speed sensors.

## Exteroceptive Sensors

An artificial hand holding a lightbulb.

Arrays of tactels can be used to provide data on what has been touched. The Shadow Hand uses an array of 34 tactels arranged beneath its polyurethane skin on each finger tip. Tactile sensors also provide information about forces and torques transferred between the robot and other objects.

Vision refers to processing data from any modality which uses the electromagnetic spectrum to produce an image. In humanoid robots it is used to recognize objects and determine their properties. Vision sensors work most similarly to the eyes of human beings. Most humanoid robots use CCDcameras as vision sensors.

Sound sensors allow humanoid robots to hear speech and environmental sounds, and perform as the ears of the human being. Microphones are usually used for this task.

## Actuators

Actuators are the motors responsible for motion in the robot.

Humanoid robots are constructed in such a way that they mimic the human body, so they use actuators that perform like muscles and joints, though with a different structure. To achieve the same effect as human motion, humanoid robots use mainly rotary actuators. They can be either electric, pneumatic, hydraulic, piezoelectric or ultrasonic.

Hydraulic and electric actuators have a very rigid behavior and can only be made to act in a compliant manner through the use of relatively complex feedback control strategies. While electric coreless motor actuators are better suited for high speed and low load applications, hydraulic ones operate well at low speed and high load applications.

Piezoelectric actuators generate a small movement with a high force capability when voltage is applied. They can be used for ultra-precise positioning and for generating and handling high forces or pressures in static or dynamic situations.

Ultrasonic actuators are designed to produce movements in a micrometer order at ultrasonic frequencies (over 20 kHz). They are useful for controlling vibration, positioning applications and quick switching.

Pneumatic actuators operate on the basis of gas compressibility. As they are inflated, they expand along the axis, and as they deflate, they contract. If one end is fixed, the other will move in a linear trajectory. These actuators are intended for low speed and low/medium load applications. Between pneumatic actuators there are: cylinders, bellows, pneumatic engines, pneumatic stepper motors and pneumatic artificial muscles.

## Planning and Control

In planning and control, the essential difference between humanoids and other kinds of robots (like industrial ones) is that the movement of the robot has to be human-like, using legged locomotion, especially biped gait. The ideal planning for humanoid movements during normal walking should result in minimum energy consumption, as it does in the human body. For this reason, studies on dynamics and control of these kinds of structures has become increasingly important.

The question of walking biped robots stabilization on the surface is of great importance. Maintenance of the robot's gravity center over the center of bearing area for providing a stable position can be chosen as a goal of control.

To maintain dynamic balance during the walk, a robot needs information about contact force and its current and desired motion. The solution to this problem relies on a major concept, the Zero Moment Point (ZMP).

Another characteristic of humanoid robots is that they move, gather information (using sensors) on the "real world" and interact with it. They don't stay still like factory manipulators and other robots that work in highly structured environments. To allow humanoids to move in complex environments, planning and control must focus on self-collision detection, path planning and obstacle avoidance.

Humanoid robots do not yet have some features of the human body. They include structures with variable flexibility, which provide safety (to the robot itself and to the people), and redundancy of movements, i.e. more degrees of freedom and therefore wide task availability. Although these characteristics are desirable to humanoid robots, they will bring more complexity and new problems to planning and control. The field of whole-body control deals with these issues and addresses the proper coordination of numerous degrees of freedom, e.g. to realize several control tasks simultaneously while following a given order of priority.

# MEDICAL ROBOT

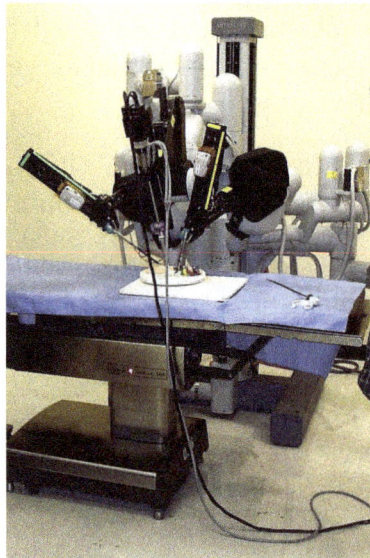

A laparoscopic robotic surgery machine. Patient-side cart of the da Vinci surgical system.

A medical robot is a robot used in the medical sciences. They include surgical robots. These are in most telemanipulators, which use the surgeon's actions on one side to control the "effector" on the other side.

## Types

- Surgical robots: either allow surgical operations to be carried out with greater precision than an unaided human surgeon, or allow remote surgery where a human surgeon is not physically present with the patient.

- Rehabilitation robots: facilitate and support the lives of infirm, elderly people, or those with dysfunction of body parts effecting movement. These robots are also used for rehabilitation and related procedures, such as training and therapy.

- Biorobots: a group of robots designed to imitate the cognition of humans and animals.

- Telepresence robots: allow off-site medical professionals to move, look around, communicate, and participate from remote locations.

- Pharmacy automation: robotic systems to dispense oral solids in a retail pharmacy setting or preparing sterile IV admixtures in a hospital pharmacy setting.

- Companion robot: has the capability to engage emotionally with users keeping them company and alerting if there is a problem with their health.

- Disinfection robot: has the capability to disinfect a whole room in mere minutes, generally using pulsed ultraviolet light. They are being used to fight Ebola virus disease.

# ENTERTAINMENT ROBOT

An entertainment robot is, as the name indicates, a robot that is not made for utilitarian use, as in production or domestic services, but for the sole subjective pleasure of the human. It serves, usually the owner or his housemates, guests or clients. Robotics technologies are applied in many areas of culture and entertainment.

Expensive robotics are applied to the creation of narrative environments in commercial venues where servo motors, pneumatics and hydraulic actuators are used to create movement with often preprogrammed responsive behaviors such as in Disneyland's haunted house ride.

Entertainment robots can also be seen in the context of media arts where artist have been employing advanced technologies to create environments and artistic expression also utilizing the actuators and sensor to allow their robots to react and change in relation to viewers.

## Toy Robot

Relatively cheap, mass-produced entertainment robots are used as mechanical, sometimes interactive, toys which perform various tasks and tricks on command. The first commercial hit was, not surprisingly, modelled on the most popular pet: the canine.

## Robotic Dog

Robot dogs as a fad have been produced with relatively little variation. These are some commercial models:

- *Teksta* a toy robot dog popular in the 1990s which was intended to be able to perform card tricks and respond to commands.

- *Aibo* (The Sony robot dog).

- *Poo-Chi*.

- *Bo-wow*.

- *I-Cybie*.

- *iDog* (Sega's robot iPod music speaker).

- *Gupi*, a robotic guinea pig.

- *Space Dog*, the remote control dog .

Robot dogs also appear fairly frequently in fiction compared to other forms of personal entertainment robots.

- *K-9* The Doctor's portable computer and robot, from the British BBC Television series *Doctor Who*.

- *Preston* - Wendolene's robot dog from the 1995 animated Wallace and Gromit film *A Close Shave*.

- *Goddard* pet of Jimmy Neutron.

## Humanoid Entertainment Robots

Despite those humanoid robots for utilitarian uses, there are some humanoid robots which aims at entertainment uses, such as Sony's QRIO and Wow Wee's RoboSapien. They are usually capable of some advanced features like Voice Recognition or Walking.

## Substitute Pets

While primitive robot toy models only execute standardized pre-programmed routines, sometimes little more than a wind-up toy could do, advancing technology allows for interaction with the user and/or other environmental stimuli (e.g. sensor-detected obstacles), thus somewhat resembling a live playmate.

Nevertheless, in the mind of some users the things can hold the loved place of a pet, as demonstrated by the fact that some even sleep with a metallic one instead of a plush cuddly toy.

In fact manufacturers even found it pays to produce a toy that is essentially designed to be nurtured, rather like an egg in some 'parenting experience simulations', as proven by the success of the Japanese Tamagotchi.

Entertainment robots can take the form of interactive communications marketing tools at trade shows wherein promotional robots move about a trade show floor providing tongue-in-cheek interaction with trade show attendees in order to bring said attendees to a particular companies trade show booth. These promotional robots are hired by companies to entertain the inform trade show attendees about products ad services available at trade shows, example www.entertainmentrobots.com rents trade show robots.

## Commercial Show Robots

As usual in the entertainment industry, capital and creativity are invested to try and top anything the private person can afford. In fact, from their owner's point of view this is a professional use, but the product is designed with as end use in mind its appreciation by the public.

Thus expensive robots are made for use as:

- Marketing tool - logically showed off by the manufacturers, to promote their products and technology, occasionally used in other promotional productions.

- Prop - inanimate performer or even artificial actor in show, TV and movie production; as technology advances, some advanced robots can, often helped with other special effects, to make them seem what cannot (yet), even be significantly more than a cast extra, such as the droids R2-D2 and C-3PO in the *Star Wars* double trilogy which have proved rather popular from the start.

## Non-commercial Art Robots

In 1956, Nicolas Schöffer created *Cysp 1 (Spatiodynamique Cybernétique)*, a robot and dancer working together to create an abstract sculpture and choreography with concrete music by Pierre Henry. These works could react to color, sound and light.

Survival Research Laboratories, in San Francisco, California, creates large destructive robotic performances to roast contemporary culture and express their distaste for the military-industrial complex.

Emergent Systems is creating large-scale interactive art environments where robots are able to respond to humans and each other as they react and evolve in the robotic installations. Autopoiesis was one such artificial life work that allowed a series of robots constructed of grapevines to both act as individuals and a group. Augmented Fish Reality allowed Siamese fighting fish to control their robots to meet across the gap of their glass fish bowls.

Intel Museum hosts the A.I. driven interactive robot, ARTI, which is short for "artificial intelligence". This robot is considered to be a work of fine art and is capable of recognizing faces, understands speech and even teaches the museum guests about the history of the museum and its founders, Robert Noyes and Gordon Moore. ARTI's face is made out of an inanimate silicon wafer.

# SERVICE ROBOTS

Service robots assist human beings, typically by performing a job that is dirty, dull, distant, dangerous or repetitive, including household chores. They typically are autonomous and/or operated by a built-in control system, with manual override options. The term "service robot" does not have a strict technical definition. The International Organization for Standardization defines a "service robot" as a robot "that performs useful tasks for humans or equipment excluding industrial automation applications".

According to ISO 8373 robots require "a degree of autonomy", which is the "ability to perform intended tasks based on current state and sensing, without human intervention". For service robots this ranges from partial autonomy - including human robot interaction - to full autonomy - without active human robot intervention. The International Federation of Robotics (IFR) statistics for service robots therefore include systems based on some degree of human robot interaction or even full tele-operation as well as fully autonomous systems.

Service robots are categorized according to personal or professional use. They have many forms and structures as well as application areas.

## Types

The possible applications of robots to assist in human chores is widespread. At present there are a few main categories that these robots fall into.

## Industrial

Industrial service robots can be used to carry out simple tasks, such as examining welding, as well as more complex, harsh-environment tasks, such as aiding in the dismantling of nuclear power stations. Industrial robots have been defined by the International Federation of Robotics as "an automatically controlled, reprogrammable, multipurpose manipulator programmable in three or more axes, which may be either fixed in place or mobile for use in industrial automation applications".

## Frontline Service Robots

Service robots are system-based autonomous and adaptable interfaces that interact, communicate and deliver service to an organization's customers.

## Domestic

The Roomba vacuum cleaner is one of the most popular domestic service robots.

Domestic robots perform tasks that humans regularly perform in non-industrial environments, like people's homes such as for cleaning floors, mowing the lawn and pool maintenance. People with disabilities, as well as people who are older, may soon be able to use service robots to help them live independently. It is also possible to use certain robots as assistants or butlers.

## Scientific

Robotic systems perform many functions such as repetitive tasks performed in research. These range from the multiple repetitive tasks made by gene samplers and sequencers, to systems which can almost replace the scientist in designing and running experiments, analysing data and even forming hypotheses. The ADAM at the University of Aberystwyth in Wales can "[make] logical assumptions based on information programmed into it about yeast metabolism and the way proteins and genes work in other species. It then set about proving that its predictions were correct."

Autonomous scientific robots perform tasks which humans would find difficult or impossible, from the deep sea to outer space. The Woods Hole Sentry can descend to 4,500 metres and allows a higher payload as it does not need a support ship or the oxygen and other facilities demanded by human piloted vessels. Robots in space include the Mars rovers which could carry out sampling and photography in the harsh environment of the atmosphere on Mars.

## Event Robots

Event Robots are starting to be used within the realms of service Robots to engage with clients and event attendees. Robots provide a great source of interaction. "Eva" photography Robot is a great example of how a Robot can be used for events to engage attendees.

## Domestic Robots

First generation Roomba vacuums the carpets in a domestic environment.

A domestic robot is a type of service robot, an autonomous robot that is primarily used for household chores, but may also be used for education, entertainment or therapy. Thus far, there are only a few limited models, though speculators, such as Bill Gates, have suggested that they could become more common in the future. While most domestic robots are simplistic, some are connected to WiFi home networks or smart environments and are autonomous to a high degree. There were an estimated 3,540,000 service robots in use in 2006[*clarification needed*], compared with an estimated 950,000 industrial robots.

## Indoor Robots

This type of domestic robot does chores around and inside homes. Different kinds include:

Robotic vacuum cleaners and floor-washing robots that clean floors with sweeping and wet mopping functions. Some use Swiffer or other disposable cleaning cloths to dry-sweep, or reusable microfiber cloths to wet-mop.

Within the ironing robots, Dressman is a mannequin to dry and iron shirts using hot air. Other ones also includes mannequin for down parts (pants, trousers and skirts). More advanced ones fold and organizes the clothes, as Laundroid (using image analysis and artificial intelligence), Effie (irons 12 items of clothing at once) and FoldiMate.

Cat litter robots are automatic self-cleaning litter boxes that filter clumps out into a built-in waste receptacle that can be lined with an ordinary plastic bag.

Robotic kitchens include Rotimatic (which makes rotis, tortillas, puris out of flour in just a few minutes), Moley Robotics MK1 and Prometheus delta robot.

Security robots such as Knightscope have a night-vision-capable wide-angle camera that detects movements and intruders. It can patrol places and shoot video of suspicious activities, too, and send alerts via email or text message; the stored history of past alerts and videos are accessible via the Web. The robot can also be configured to go into action at any time of the day.

Atlas is a robot built to perform in house task such as sweeping, opening doors, climbing stairs, etc.

## Outdoor Robots

A robotic lawn mower is a lawn mower that is able to mow a lawn by itself after being programmed. Once programmed, this invention repeats the operation by itself according to its programming. Robotic lawn mowers come with a power unit which may be an electric motor or internal combustion engine. This provides power to the robot and allows it to move itself and its cutting blades. There is also a control unit which helps the mower move. This unit also contains a memory unit which records and memorizes its operation programming. Its memorized route includes the length of travel in a given direction and turns angles. This allows the same lawn to be mowed repeatedly without having to reprogram. The steering unit acquires an operation signal and propels the lead wheel, which leads the mower, go guide along the programmed route.

Some models can mow complicated and uneven lawns that are up to three-quarters of an acre in size. Others can mow a lawn as large as 40,000 square feet (3,700 m2), can handle a hill inclined up to 27 degrees.

There are also automated pool cleaners that clean and maintain swimming pools autonomously by scrubbing in-ground pools from the floor to the waterline in 3 hours, cleaning and circulating more than 70 US gallons (260 l) of water per minute, and removing debris as small as 2 μm in size.

Gutter-cleaning robots such as Looj use brushes and rubber blades to remove debris from rain gutters; users operate the device using a remote.

Window cleaning robots are most commonly used to clean outdoor windows, more specifically house windows. However, it may be used on other types of windows, such as ones on tall buildings and structures. This robot contains a movement system which allows the robot to navigate itself across the window surface in a defined direction. It also has a powered agitator located by the cleaning pad. When activated, the agitator gets rid of debris and dirt from the window surface. The cleaning pad directly interacts with the window surface and is directly responsible for removing the dirt by filling itself with specialized window cleaning fluid.

A window-washing robot commonly uses two magnetic modules to navigate windows as it sprays the cleaning solution onto microfiber pads to wash them. It covers about 1,601 square feet (148.7 m²) per charge.

## Toys

Robotic toys, such as the well known Furby, have been popular since 1998. There are also small humanoid remote-controlled robots as well as electronic pets, such as robotic dogs. They have also have been used by many universities in competitions such as the RoboCup.

## Social Robots

Social robots take on the function of social communication. Domestic humanoid robots are used by elderly and immobilized residents to keep them company.

Home-telepresence robots can move around in a remote location and let one communicate with people there via its camera, speaker, and microphone.

Network robots link ubiquitous networks with robots, contributing to the creation of new lifestyles and solutions to address a variety of social problems including the aging of population and nursing care. Robots built for therapy have been in production for quite some time now. Some of these uses can be for autism or physical therapy.

## Industrial Robots

An industrial robot can be defined as a robot system that is used for manufacturing. A wide range of tasks can be performed by the industrial manufacturing robots as they are included with diverse capabilities.

Industrial robots perform various tasks (such as picking up and clamping activities, spot and arc welding, clamping for machining, manipulation and transfer of parts) with the help of sensors, computer software, and a network of complex mechanical gestures.

Furthermore, industrial robots are automated, programmable and capable of movement in two or more axes. Applications of robots include welding, painting, assembly, pick and place for print-ed circuit boards, packaging and labeling, palletizing, product inspection, and testing; all accomplished with high endurance, speed, and precision.

- Axis – Axis can be defined as a direction that is used to state the motion of a robot in a linear or rotary mode

- Speed – In this case, speed can be defined as the speed of a robot used to position the end of its arm when all axes are moving.

- Acceleration – How quickly a robotic arm can pick up the pace

- Accuracy – Accuracy can be defined as exactness of a robot on how closely it can reach a commanded position.

- Repeatability – How well the robot will return to a programmed position.

- Carrying capacity or payload – how much weight a robot can lift

- Prismatic Joint: It provides a linear sliding movement between two bodies (it is often called as a slider). It can be formed with a polygonal cross-section to resist rotation.

- Kinematics – The actual arrangement of rigid members and joints in the robot, which determines the robot's possible motions. Classes of robot kinematics include articulated, Cartesian, parallel and SCARA.

## Types of Industrial Robots

There are different types of industrial robots based on specifications and applications. Various types of industrial robots include non-servo robots, servo robots, programmable robots, and computer programmable robots.

- Non-servo robots: These robots are used to move and place objects. That means, these robots will be capable to pick up an object and transport the object, place it down.

- Servo robots: Servo robots include manipulators, effectors, robotic appendages that function as the arms and hands of the robot.

- Programmable robots: These robots store commands in a database i.e. they can repeat a task a pre-determined number of times.

- Computer programmable robots: These robots are essentially servo robots that can be controlled remotely, via a computer.

## Classification by Types of Robots

Robots are mostly classified into five major configurations based on their mechanical structure. They are:

- Cartesian robot: Cartesian robot consists of three prismatic joints and axes are concurrent with a Cartesian coordinate system

- SCARA robot: SCARA (Selective Compliance Assembly Robot Arm) robot is included with two parallel rotary joints to provide compliance in a plane.

- Articulated robot: Articulated robot is a robot whose arm has at least three rotary joints

- Parallel robot: A parallel robot is a robot whose arms have concurrent prismatic or rotary joints

Let us check out the classification of robots in detail:

## Cartesian Robot

A Cartesian robot can be defined as an industrial robot whose three principal axes of control are linear and are at right angles to each other.

Using their rigid structure, they can carry high payloads. They can perform some functions such as pick and place, loading and unloading, material handling, and so on. Cartesian robots are also called as Gantry robots as their horizontal member supports both the ends.

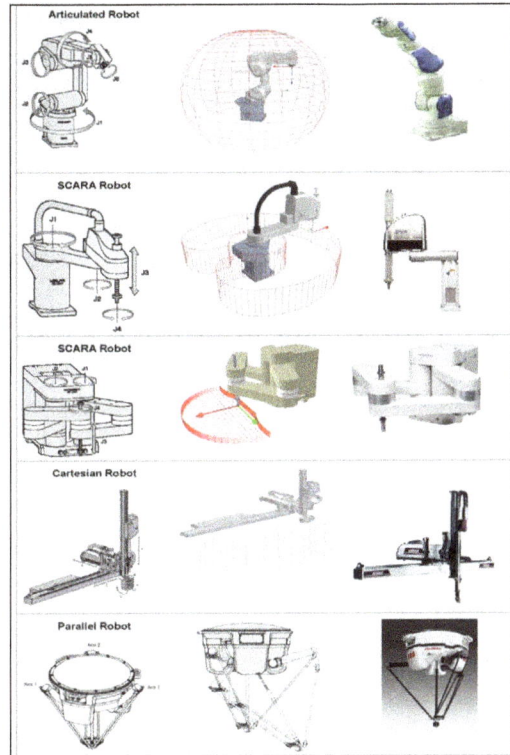

Classification of industrial robots by mechanical structure.

Cartesian robots are also known as linear robots or XYZ robots as they are outfitted with three rotary joints for assembling XYZ axes.

## Examples of Linear/Cartesian/Gantry Robots

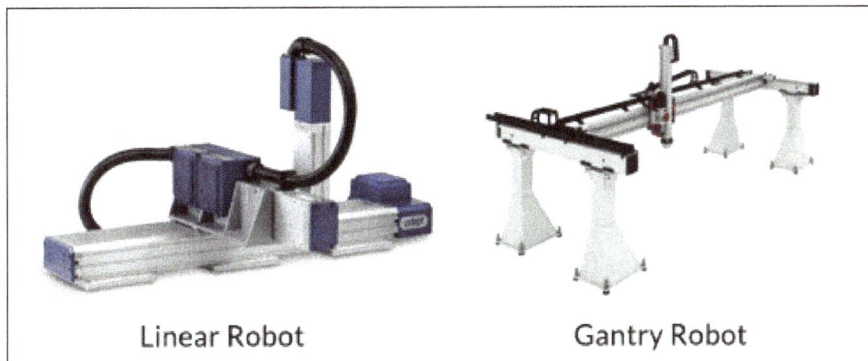

Linear Robot                    Gantry Robot

## Applications

Cartesian robots can be used in sealing, handling for plastic moulding, 3D printing, and in a computer numerical control machine (CNC). Pick and place machines and plotters work on the principle of the Cartesian robots. They can handle heavy loads with high positioning accuracy.

## Examples of Applications of Linear/Cartesian/Gantry Robots

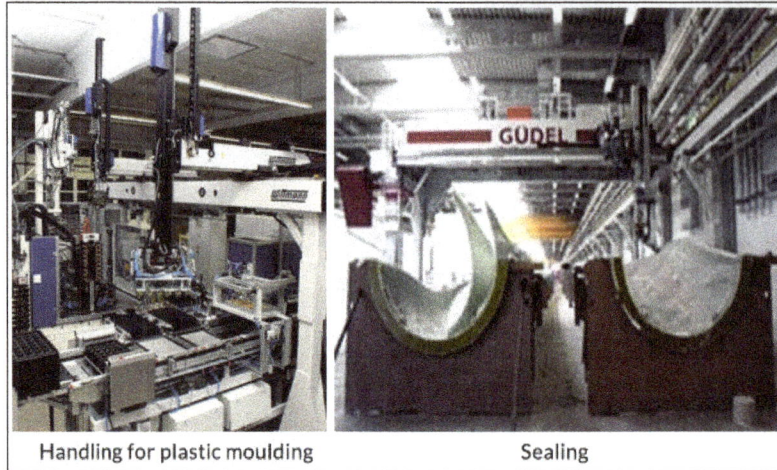

| Handling for plastic moulding | Sealing |

## Advantages

- Highly accurate & speed.
- Less cost.
- Simple operating procedures.
- High payloads.
- Very versatile working.
- Simplifies robot and master control systems.

## Disadvantages

- They require a large volume of space to operate.

## SCARA Robot

The SCARA acronym stands for Selective Compliance Assembly Robot Arm or Selective Compliance Articulated Robot Arm.

The robot was developed under the guidance of Hiroshi Makino, a professor at the University of Yamanashi. The arms of SCARA are flexible in the XY-axes and rigid in the Z-axis that makes it to familiarize to holes in the XY-axes.

In the X-Y direction, the arm of SCARA robot will be compliant and strong in the 'Z' direction due to the virtue of the SCARA's parallel-axis joint layout. Hence the term, Selective Compliant.

This robot is used for various types of assembly operations i.e. a round pin can be inserted in a round hole without binding using this. These robots are faster and cleaner than comparable robot systems and they are based on serial architectures, that means the first motor should carry all other motors.

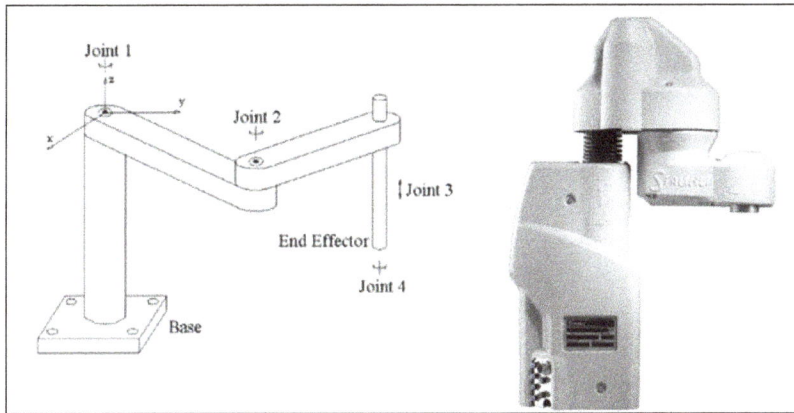

Representation of XYZ axes of a SCARA robot.

## Examples of SCARA Robots

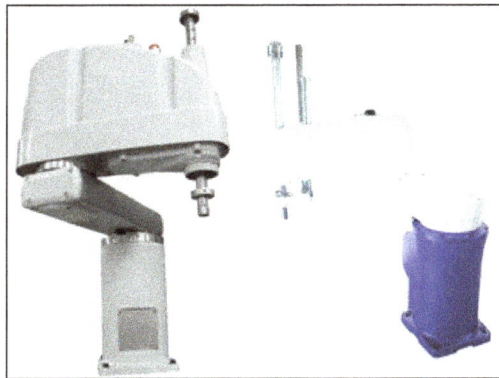

## Applications

SCARA robots are used for assembly, packaging, palletisation, and machine loading.

## Examples of Applications of SCARA Robots

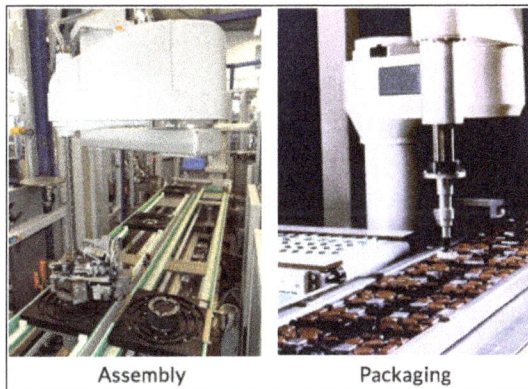

Assembly          Packaging

## Advantages

- High speed capabilities.

- Perform great in short-stroke, fast assembly and pick-and-place applications.
- It contains Donut Shaped work envelope.

## Disadvantages

SCARA robot typically requires dedicated robot controller in addition to line master controller like PLC/PC.

## Articulated Robot

An Articulated robot can be defined as a robot with rotary joint and these robots can range from simple two-jointed structures to systems with 10 or more interacting joints.

These robots can reach any point as they work in three dimensional spaces. On the other hand, articulated robot joints can be parallel or orthogonal to each other with some pairs of joints parallel and others orthogonal to each other. As articulated robots have three revolute joints, the structure of these robots is very similar to the human arm.

## Examples of Articulated Robots

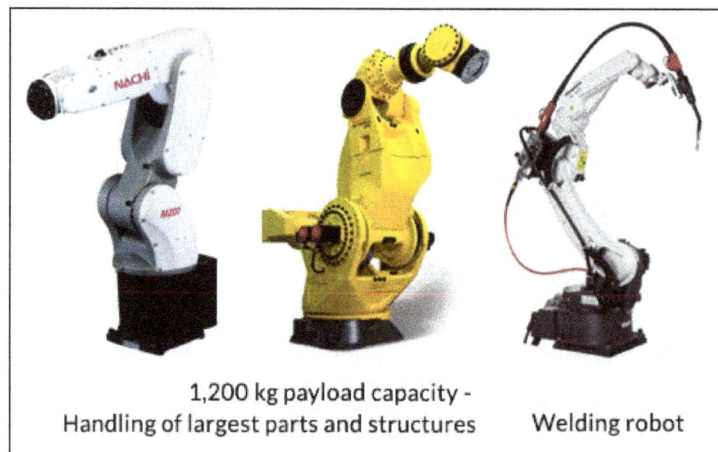

1,200 kg payload capacity -
Handling of largest parts and structures          Welding robot

## Applications

Articulated robots can be used in Robots palletizing food (Bakery), Manufacturing of steel bridges, cutting steel, Flat-glass handling, heavy duty robot with 500 kg payload, Automation in foundry industry, heat resistant robot, metal casting, and Spot Welding.

## Examples of Applications of Articulated Robots

## Advantages

- High speed.
- Large working envelope.
- Great in unique controller, welding and painting applications.

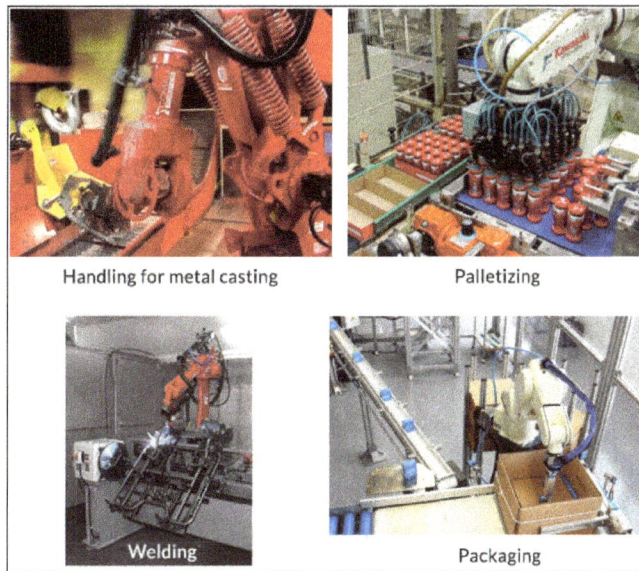

Handling for metal casting

Palletizing

Welding

Packaging

## Disadvantage

Typically requires dedicated robot controller in addition to line master controller like PLC/PC.

## Parallel Robots

Parallel robots are also known as parallel manipulators or generalized Stewart platforms.

A parallel robot is a mechanical system that uses several computer-controlled serial chains to support a single platform, or end-effector.

Furthermore, a parallel robot can be formed from six linear actuators that maintain a movable base for devices such as flight simulators. These robots prevent redundant movements and to carry out this mechanism, their chain is designed to be short, simple.

They are known as:

- High speed and high precision milling machines.
- Micro manipulators mounted on the end effector of larger but slower serial manipulators.
- Examples of parallel robots.

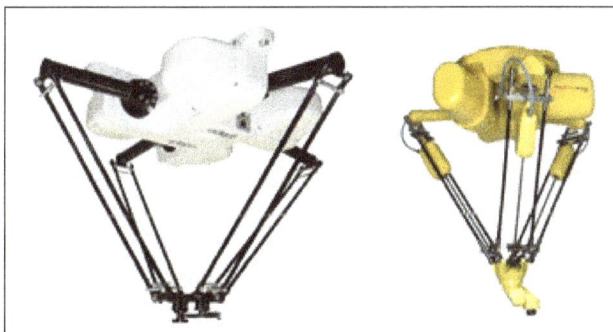

## Applications

Parallel robots are used in various industrial applications such as:

- Flight simulators.

- Automobile simulators.

- In work processes.

- Photonics/optical fiber alignment.

They are used in limit in the workspaces. To perform a desired manipulation, it would be very difficult and can lead to multiple solutions. Two examples of popular parallel robots are the Stewart platform and the Delta robot.

## Examples of Applications of Parallel Robots

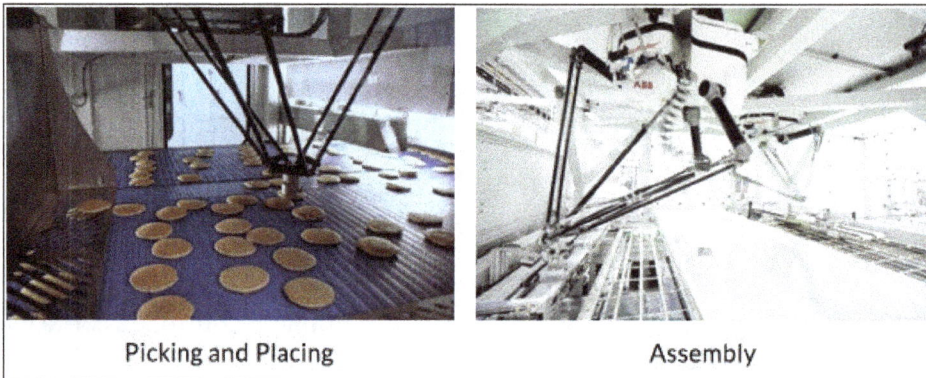

Picking and Placing                 Assembly

## Advantages

- Very high speed.

- Contact lens shaped working envelope.

- Excels in high speed, lightweight pick and place applications (candy packaging).

## Disadvantages

It requires dedicated robot controller in addition to line master controller like PLC/PCs.

## Programming of Robots to Perform a Required Position

Robots are programmed by humans to perform complicated and required tasks. Here, let us check out how robots are programmed to carry out the required position:

- Positional Commands: A robot can perform the required position using a GUI or text based commands in which the essential X-Y-Z position may be specified and edited.

- Teach Pendant: Using a teach pendant method, we can teach the positions to a robot.

Teach pendent is a handheld control and programming unit that contain the capability to manually send the robot to a desired position.

A teach pendant can be disconnected after the completion of programming. But, the robot runs the program, which was fixed in controller.

- Lead-by-the-nose: Lead-by-the-nose is a technique which will be included by many robot manufacturers. In this method, one user holds the robot's manipulator, while another person enters a command that helps to de-energize the robot which will make it to go into limp.

Then, user can move the robot to the required position (by hand) while the software records these positions into memory. Several robot manufacturers use this technique for performing paint spraying.

- Robotic Simulator: A robotic simulator helps not to depend on the physical operation of the robot arm. Following this method helps to save time in the design of robotics applications and enhances the safety level. On the other hand, programs (which are written in various programming languages) can be tested, run, taught, and debugged using the robotic simulation software.

- Machine Operator: A machine operator can be used to make adjustments within a program. These operators use touch-screen units that serve as the operator control panel.

## Serial Robots

A robot is said to be a serial robot or serial (open-loop) manipulator if its kinematic structure takes the form of an open loop-chain. It is one of the most common industrial robot.

## Serial Robot Designs

In its most general form, a serial robot design consists of a number of rigid links connected with joints. Simplicity considerations in manufacturing and control have led to robots with only revolute or prismatic joints and orthogonal, parallel and/or intersecting joint axes (instead of arbitrarily placed joint axes).

The inverse kinematics of any serial manipulator with six revolute joints, and with three consecutive joints intersecting, can be solved in closed-form, i.e., analytically.

This result had a tremendous influence on the design of industrial robots: until 1974, when Cincinnati Milacron launched its T 3 robot (which has three consecutive parallel joints, i.e., intersecting at infinity), all industrial manipulators had at least one prismatic joint while since then, most industrial robots are wrist-partitioned 6R manipulators, such as shown in figures. These 6R robots have six revolute joints, and their last three joint axes intersect orthogonally, i.e., they form a spherical wrist such as, for example, the ZXZ wrist whose motion capabilities are illustrated in Fig. Hence, they can achieve any possible orientation.

As Pieper proved, this construction leads to a decoupling of the position and orientation kinematics, for the forward as well as the inverse problems. The inverse solution for the three wrist joints is a copy of the inverse Euler angle problem; the remaining three joints are then found by solving a polynomial of, at most, fourth order, whatever their kinematic structure is. The extra structural simplifications (i.e., parallel or orthogonal axes) introduced in the serial robots of, for example, Figures, lead to even simpler solutions. (Intuitively speaking, each intelligently chosen geometric constraint imposed on the kinematic structure simplifies the calculations, because it reduces the solution search space.) The simplest kinematics are found in the SCARA (Selectively Compliant Assembly Robot Arm) design, Fig. This design has three vertical revolute joint axes, and one vertical prismatic joint at the end. SCARA robots are mainly used for "pick-and-place" operations. In such a task, the robot must be stiff in the vertical direction (because it has to push things into other things) and a bit compliant in the horizontal plane, because of the imperfect relative positioning between the manipulated object and its counterpart on the assembly table. This desired selective compliance behaviour is intrinsic to the SCARA design; hence the name of this type of robots.

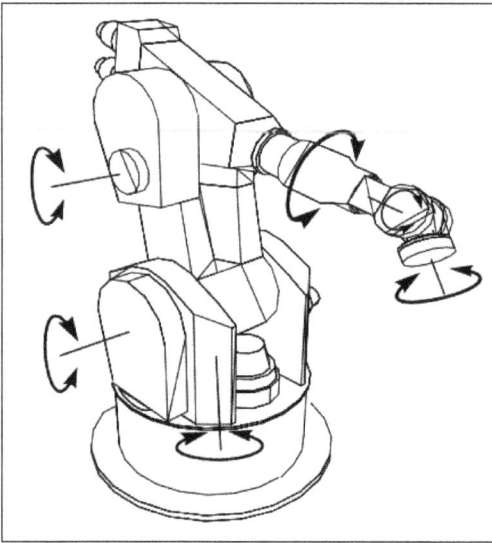

A Kuka-160 serial robot.

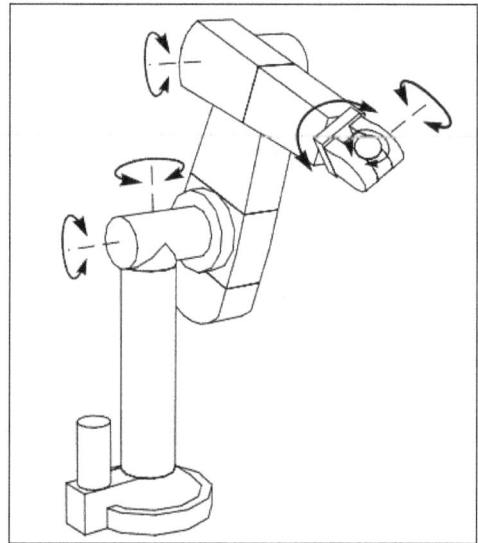

A Staubli (formerly Unimation) "PUMA" serial robot.

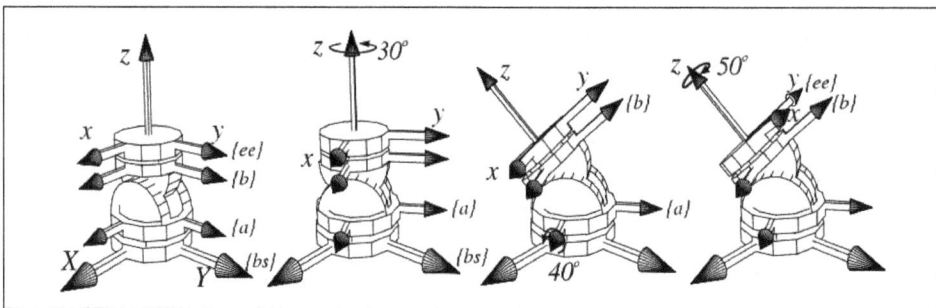

The three motion capabilities of a spherical ZXZ wrist. This terminology comes from the observation that the depicted wrist makes rotations about the Z, X and Z axes of the subsequent joints.

Hybrid designs. A last industrially important class of "serial" robot arms are the gantry robots, Figure. They have three prismatic joints to position the wrist, and three revolute joints for the wrist. Strictly speaking, a gantry robot is a combination of a parallel XYZ translation structure with a serial spherical wrist. The parallel construction is very stiff (cf. metal cutting machines) so that these robots are very accurate. In large industrial applications (such as welding of ship hulls or other large objects) a serial manipulator is often attached to a two or three degrees of freedom gantry structure, in order to combine the workspace and dexterity advantages of both kinematic structures.

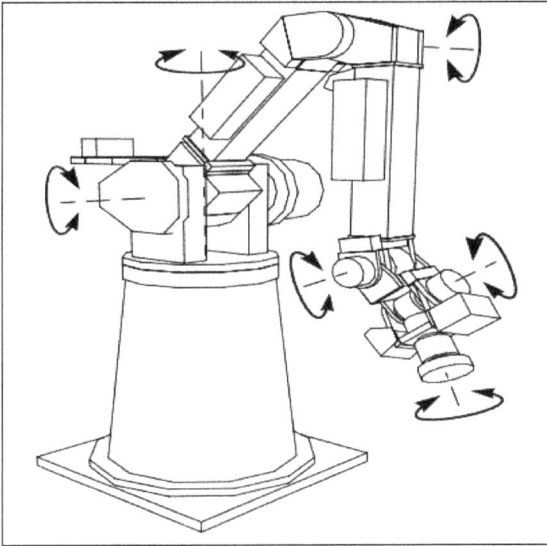

The Cincinnati Milacron T 3 serial robot.          An Adept SCARA robot.

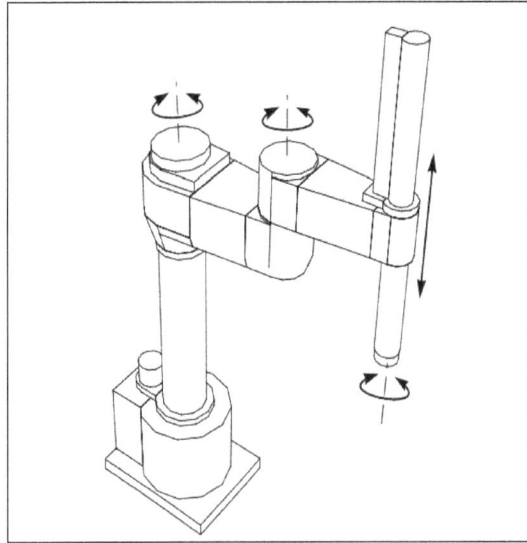

Design characteristics. The examples above illustrate the common design characteristics of (most) industrial serial robot arms:

1.  They are anthropomorphic, in the sense that they have a "shoulder," (first two joints) an "elbow," (third joint) and a "wrist" (last three joints). So, in total, they have the six degrees of freedom needed to put an object in an arbitrary position and orientation.

2.  Almost all commercial serial robot arms have only revolute joints. Compared to prismatic joints, revolute joints are cheaper and give a larger dextrous workspace for the same robot volume.

3.  Serial robots are very heavy, compared to the maximum load they can move without loosing their accuracy: their useful load to own-weight ratio is worse than 1/10! The robots are so heavy because the links must be stiff: flexible links cause deformations, and hence position and orientation errors at the end-point.

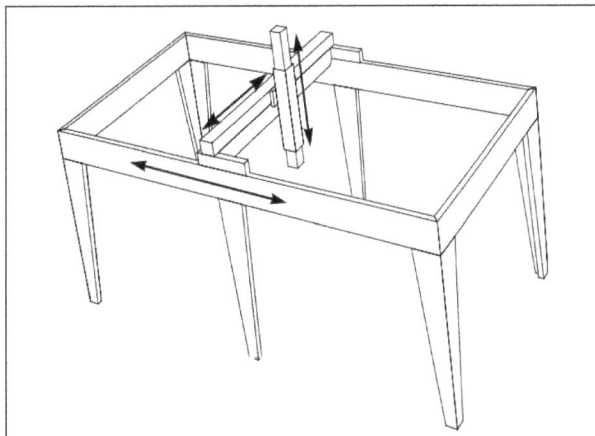

A gantry robot. (Only the first three prismatic degrees of freedom are shown.)

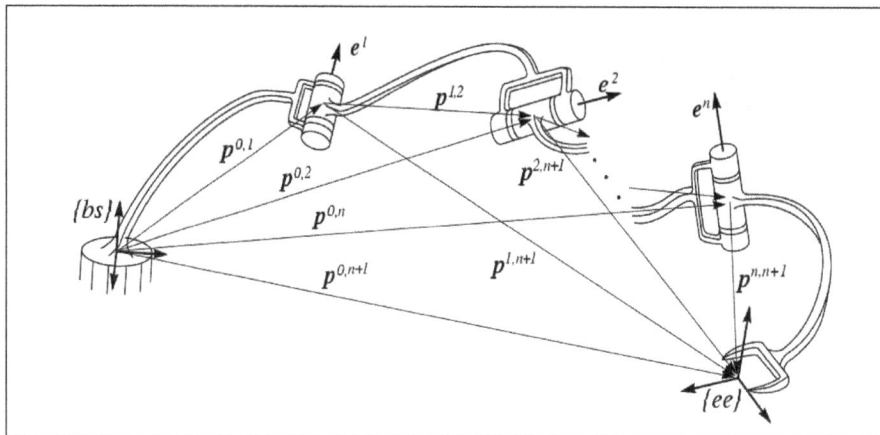

Notations used in the geometrical model of a serial kinematic chain.

4. Simplicity of the forward and inverse position and velocity kinematics has always been one of the major design criteria for industrial manipulator arms. Hence, almost all of them have a very special kinematic structure, with a majority having the 321 design of Fig. The special kinematic structures have efficient closed-form solutions because they allow for the decoupling of the position and orientation kinematics. The geometric feature that generates this decoupling is the intersection of joint axes.

## Workspace

The reachable workspace of a robot's end-effector is the manifold of reachable frames, irrespective of how the robot must move to reach those frames. The dextrous workspace is the subset of the reachable workspace where the robot can generate velocities in all directions, i.e., it can translate the manipulated object with three degrees of translation freedom, and rotate the object with three degrees of rotation freedom.

The relationships between joint space and Cartesian space coordinates of the object held by the robot are in general multiple-valued: the same pose in the reachable workspace can be reached by the serial arm in different ways, each with a different set of joint coordinates. Hence, the reachable workspace of the robot is divided in configurations (also called assembly modes), in which the kinematic relationships are locally one-to-one.

## 321 Kinematic Structure

Because all-revolute joint manipulators have good workspace properties, and because a sequence of three intersecting joint axes introduces significant simplifications in the kinematic algorithms, most industrial robot arms now have a kinematic structure as shown in figure (Vic Scheinman of Stanford University was, to the best of the authors' knowledge, the first to come up with this de-sign, but he did not write it up in any readily accessible publications) The design is an example of a 6R wrist-partitioned manipulator: the last three joint axes intersect orthogonally at one point. Moreover, the second and third joints are parallel, and orthogonal to the first joint. These facts motivate the name of "321 " robot arm: the three wrist joints intersect; the two shoulder and elbow joints are parallel, hence they intersect at infinity; the first joint orthogonally intersects the first shoulder joint.

The 321 structure can be given a link frame transformation convention that is much simpler than the Denavit-Hartenberg or Hayati-Roberts conventions: its geometry is determined by orthogonal and parallel joint axes, and by only four link lengths $l_1$, $l_2$, $l_3$ and $l_6$, because the wrist link lengths l4 and l5 are zero. In this simpler convention, the reference frames are chosen to be all parallel when the robot is in its fully upright configuration. This configuration is defined to be the kinematic zero position in the rest of this text, i.e., all joint angles are defined to be zero in this position. The six joints are defined to rotate in positive sense about, respectively, the $+Z^1, -X^2, -X^3, +Z^4, -X^5$, and $+Z^6$ axes, such that positive joint angles make the robot "bend forward" from its kinematic zero position. Many industrial robots have a 321 kinematic structure, but it is possible that the manufacturers defined different zero positions and different positive rotation directions for some joints. These differences are easily compensated by (constant) joint position offsets and joint position sign reversals.

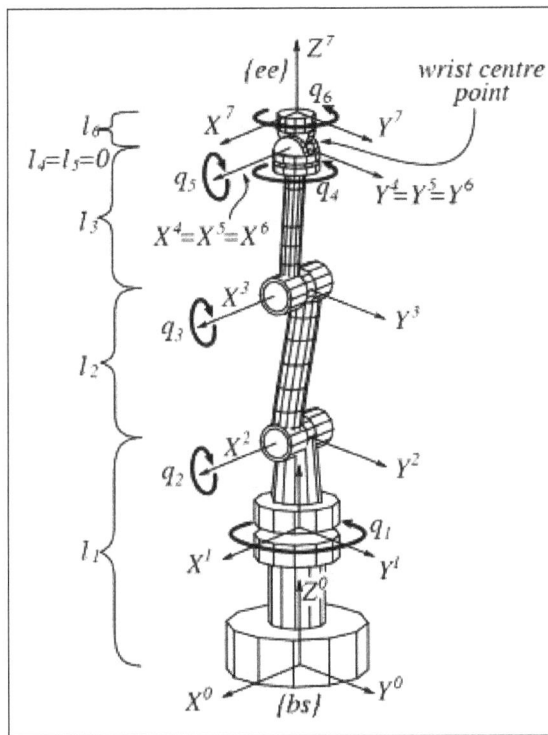

321 kinematic structure in the "zero" position: all link frames are parallel and all origins lie on the same line.

321 kinematic structure with offsets. Many other industrial robots, such as for example the PUMA, have a kinematic structure that deviates a little bit from the 321 structure of Figure :

1.  Shoulder offset: frame {3} in Figure is shifted a bit along the X-axis. This brings the elbow off-centre with respect to the line of joint 1.

2.  Elbow offset: frame {4} in Figure is shifted a bit along the Y -axis. This brings the wrist centre point off-centre with respect to the forearm.

The reasons for the offsets will become clear in the Section on singularities (Sect. 15): the offsets move the singular positions of the robot away from places in the workspace where they are likely to cause problems.

## Forward Position Kinematics

The forward position kinematics (FPK) of a kinematic chain (not just serial kinematic chains, but all kinematic chains!) with n joints solves the following problem:

Given the joint positions $q = (q_1 \ldots q_n)^T$, what is the corresponding end-effector pose?

In the case of a serial kinematic chain, the solution is always unique: one given joint position vector always corresponds to only one single end-effector pose. The following Section proves that the FK problem is not difficult to solve, even for a completely arbitrary serial kinematic structure.

## General FPK: Link Transform Algorithm

The easiest approach to calculate the FPK is to apply the composition formula for homogeneous transformation matrices from the end-effector frame {ee} to the base frame {bs}: Each link transform can be constructed easily for any chosen link frame definition; for example, from the Denavit-Hartenberg definition (although that definition has no particular advantages at all in this case). This equation can be replaced by a sequence of matrix exponentiations, in which each matrix represents a linear or angular velocity generated by one of the prismatic or revolute joints in the serial arm. Equation ${}^{ee}_{bs}T = {}^{0}_{bs}T \, {}^{1}_{0}T(q_1) \, {}^{2}_{1}T(q_2) \ldots {}^{n}_{n-1}T(q_n) \, {}^{ee}_{n}T$ is then called the product of exponentials formula for a serial robot.

$$
{}^{ee}_{bs}T = {}^{0}_{bs}T \, {}^{1}_{0}T(q_1) \, {}^{2}_{1}T(q_2) \ldots {}^{n}_{n-1}T(q_n) \, {}^{ee}_{n}T
$$

$$
{}_{4}J = \begin{pmatrix} A & B \\ C & 0_3 \end{pmatrix}
$$

$$
\partial fT(q)/\partial q_i
$$

This approach works for any serial robot, with any number of revolute and/or prismatic joints. The resulting mapping from joint angles to end-effector pose is nonlinear in the joint angles. When implementing this procedure in a computer program, one should, of course, not code the complete matrix multiplications of Eq. ${}^{ee}_{bs}T = {}^{0}_{bs}T \, {}^{1}_{0}T(q_1) \, {}^{2}_{1}T(q_2) \ldots {}^{n}_{n-1}T(q_n) \, {}^{ee}_{n}T$, since (i) the last rows of the homogeneous transformation matrices are zeros and ones, and (ii) many robots have a kinematic structure that generates many more zeros in the rest of the matrices too.

## Closed-form FPK for 321 Structure

Serial manipulators of the 321 type allow for the decoupling of the robot kinematics at the wrist, for position as well as velocity, and for the forward as well as the inverse problems. This decoupling follows from the fact that the wrist has three intersecting revolute joints, and hence any orientation can be achieved by the wrist alone. This Section (and all the following Sections that treat closed-form 321 kinematics) starts by "splitting" the manipulator at the wrist centre point. This has the following advantages:

1. The position and linear velocity of the wrist centre point are completely determined by the first three joint positions and velocities.

2. The relative orientation and angular velocity of the last wrist frame with respect to the first wrist frame are easy to find from the three wrist joint angles.

3. The relative pose of the end-effector frame with respect to the last wrist frame is a constant translation along the Z 6 -axis. A similar relationship holds between the frames on the base and on the first link.

The following procedure applies this approach to the forward position kinematics:

## Closed-form FPK

Step 1 The closed-form forward orientation kinematics of the wrist is an instantiation of a ZXZ Euler angle set, upto the small difference that the positive sense of the rotation of the second wrist joint is about the −X axis. The resulting homogeneous transformation matrix from {4} to {6}, with the angles α, β and γ replaced by the joint angles $q_4$, $q_5$ and $q_6$, and taking into account the sign difference for $q_5$, is:

$$
{}^6_4T = \begin{pmatrix} {}^6_4R & \mathbf{0}_{3\times1} \\ \mathbf{0}_{1\times3} & 1 \end{pmatrix} = \begin{pmatrix} c_6c_4 - s_6c_5s_4 & -s_6c_4 - c_6c_5s_4 & -s_5s_4 & 0 \\ c_6s_4 + s_6c_5c_4 & -s_6s_4 + c_6c_5c_4 & s_5c_4 & 0 \\ -s_6s_5 & -c_6s_5 & c_5 & 0 \\ 0 & 0 & 0 & 1 \end{pmatrix}
$$

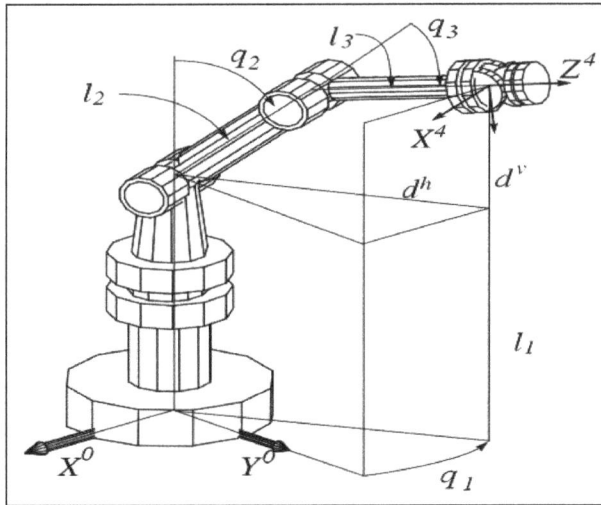

Kinematics of first three joints of the 321 manipulator.

Step 2 The pose of the wrist reference frame {4} with respect to the base referencee., ${}^4_0T$) is determined by the first three joints. $q_2$ and $q_3$ are parallel, so they move the centre of the wrist in a plane, whose rotation about the Z axis of the base reference frame {bs} is determined by q1 only. $q_2$ and $q_3$ move the wrist to a vertical height d v above the shoulder reference frame {2} (i.e., $d^v + l_1$ above $X^0Y^0$) and to a horizontal distance d h in the arm plane, i.e., the Y Z-plane of {2}:

$$ d^v = c_2l_2 + c_{23}l_3, \qquad d^h = s_2l_2 + s_{23}l_3, $$

with $c_2 = \cos(q_2)$, $c_{23} = \cos(q_2 + q_3)$, etc. The contribution of the first three joints to the total orientation matrix consists of a rotation about $Z_1$, over an angle $q_1$, followed by a rotation about the moved $X_2$-axis, over an angle $q_2 + q_3$:

$$
{}^4_0\boldsymbol{R} = \begin{pmatrix} c_1 & -s_1 & 0 \\ s_1 & c_1 & 0 \\ 0 & 0 & 1 \end{pmatrix} \begin{pmatrix} 1 & 0 & 0 \\ 0 & c_{23} & -s_{23} \\ 0 & s_{23} & c_{23} \end{pmatrix} = \begin{pmatrix} c_1 & -s_1 c_{23} & s_1 s_{23} \\ s_1 & c_1 c_{23} & -c_1 s_{23} \\ 0 & s_{23} & c_{23} \end{pmatrix}.
$$

Step 3 The pose of the end-effector reference frame $\{7\} = \{ee\}$ with respect to the last wrist reference frame $\{6\}$ (i.e., $^7_6 T$) corresponds to a translation along $Z_6$ over a distance $l_6$:

$$
{}^7_6\boldsymbol{T} = \begin{pmatrix} {}^7_6\boldsymbol{R} & l_6 \boldsymbol{e}_z \\ \boldsymbol{0}_{1\times3} & 1 \end{pmatrix} = \begin{pmatrix} 1 & 0 & 0 & 0 \\ 0 & 1 & 0 & 0 \\ 0 & 0 & 1 & l_6 \\ 0 & 0 & 0 & 1 \end{pmatrix}.
$$

Step 4 Hence, the total orientation ${}^{ee}_{bs}R$ follows from equations:

$$
{}^6_4\boldsymbol{T} = \begin{pmatrix} {}^6_4\boldsymbol{R} & \boldsymbol{0}_{3\times1} \\ \boldsymbol{0}_{1\times3} & 1 \end{pmatrix} = \begin{pmatrix} c_6 c_4 - s_6 c_5 s_4 & -s_6 c_4 - c_6 c_5 s_4 & -s_5 s_4 & 0 \\ c_6 s_4 + s_6 c_5 c_4 & -s_6 s_4 + c_6 c_5 c_4 & s_5 c_4 & 0 \\ -s_6 s_5 & -c_6 s_5 & c_5 & 0 \\ 0 & 0 & 0 & 1 \end{pmatrix},
$$

$$
{}^4_0\boldsymbol{R} = \begin{pmatrix} c_1 & -s_1 & 0 \\ s_1 & c_1 & 0 \\ 0 & 0 & 1 \end{pmatrix} \begin{pmatrix} 1 & 0 & 0 \\ 0 & c_{23} & -s_{23} \\ 0 & s_{23} & c_{23} \end{pmatrix} = \begin{pmatrix} c_1 & -s_1 c_{23} & s_1 s_{23} \\ s_1 & c_1 c_{23} & -c_1 s_{23} \\ 0 & s_{23} & c_{23} \end{pmatrix}. \text{ and}
$$

$$
{}^7_6\boldsymbol{T} = \begin{pmatrix} {}^7_6\boldsymbol{R} & l_6 \boldsymbol{e}_z \\ \boldsymbol{0}_{1\times3} & 1 \end{pmatrix} = \begin{pmatrix} 1 & 0 & 0 & 0 \\ 0 & 1 & 0 & 0 \\ 0 & 0 & 1 & l_6 \\ 0 & 0 & 0 & 1 \end{pmatrix}. :
$$

$$
{}^{ee}_{bs}\boldsymbol{R} = {}^7_0\boldsymbol{R} = {}^4_0\boldsymbol{R} \, {}^6_4\boldsymbol{R} \, {}^7_6\boldsymbol{R}.
$$

Step 5 The position of the wrist centre (i.e., the origin of $\{4\}$ with respect to the base $\{0\}$) is:

$$
{}_{bs}\boldsymbol{p}^{wr} = {}_0\boldsymbol{p}^{wr} = \begin{pmatrix} c_1 d^h \\ s_1 d^h \\ l_1 + d^v \end{pmatrix},
$$

and the position of the end-effector (i.e., the origin of $\{ee\}$ with respect to the base $\{0\}$) is:

$$
{}_{bs}\boldsymbol{p}^{ee} = {}_{bs}\boldsymbol{p}^{wr} + {}^{ee}_{bs}\boldsymbol{R} \left(0\,0\,l_6\right)^T.
$$

Step 6 Equations (6) and (8) yield the final result:

$$\,_{bs}^{ee}T = \left( \begin{array}{cc} \,_{bs}^{ee}R & \,_{bs}pe^e \\ \mathbf{0}_{1\times3} & 1 \end{array} \right).$$

## Accuracy, Repeatability and Calibration

Finding the pose of the end-effector given the actual joint positions relies on a mathematical ideal-ization: in reality, the mathematical model is not 100% accurate (due to manufacturing tolerances) and the joint positions are not measured with infinite accuracy. This means that the real pose dif-fers from the modelled pose. The smaller this difference, the better the absolute (positioning) ac-curacy of the robot, i.e., the mean difference between the actual pose and the pose calculated from the mathematical model. Absolute accuracy, however, is not the only important factor: in most industrial applications, robots are programmed on line, i.e, a human operator moves the end-ef-fector to the desired pose and then stores the current values of the joint positions in the robot's electronic memory. This way, the absolute accuracy is not relevant, but rather the repeatability of the robot, i.e., the (mean) difference between the actual poses attained by the robot in subsequent (identical) motions to the same desired pose, whose corresponding joint values have been stored in memory.

The robot's repeatability is much better than its absolute accuracy, typically an order of magni-tude. For good industrial robots, the repeatability is of the order of 0.1mm. This is the static repeat-ability, i.e., the robot moves to the desired pose, and comes to a halt while the robot controller has sufficient time to make the robot reach this pose as accurately as possible.

Off-line programming and calibration More and more robots are programmed off line. This means that CAD (Computer Aided Design) drawings of the robot and its environment are used to (i) first interactively program the robot task on a graphical workstation until the desired functionality is reached, and (ii) then download the final task program to the robot work-cell. This approach has the advantage that it does not occupy the work-cell during the programming phase; its disadvan-tage is that it applies only to workcells in which the robots have a (very) high absolute accuracy, and the robot's environment is known with the same accuracy. Since it is expensive to build robots that correspond exactly to their nominal geometrical models, the practical solution to the absolute accuracy problem is to calibrate the robot, i.e., to adapt the geometrical model to the real kine-matic structure of the robot, before bringing the robot in operation. ("Intelligent" robots follow an alternative approach: they use sensors to detect the errors on line and adapt the robot task accord-ingly.) A typical calibration procedure looks like this.

## Calibration Algorithm

Step 1 (Error model). One starts from the nominal geometric robot model, and adds a set {P} of n error parameters. These parameters model the possible geometrical differences between the nom-inal and real kinematic structures. Of course, such an error model is a practical trade-off between (i) accuracy, and (ii) complexity. Common error parameters are offsets on the joint positions, joint axis line parameters, and base and endeffector frames. For example, $\Delta\alpha$, $\Delta h$, $\Delta\theta$, $\Delta d$ in the De-navit-Hartenberg link frame convention.

Step 2 (Data collection). The robot is moved to a large set of N different poses where its end-effector homogeneous transform ${}^{ee}_{bs}T_m(q_i), i = 1, \ldots, N$ is calculated from the measured joint values and the nominal kinematic model. The number N of sampled poses is much larger than the number n of error parameters. The real end-effector and/or link frame poses are measured with an accurate 3D measurement device (e.g., based on triangulation with laser or visual pointing systems).

Step 3 (Parameter fitting). The real poses are expressed as a Taylor series in the error parameters $P_j, j = 1, \ldots, n$:

$$ {}^{ee}_{bs}T(q_i, P) = {}^{ee}_{bs}T(q_i, 0) + \sum_{j=1}^{n} \left\{ \partial\left( {}^{ee}_{bs}T(q_i, P)\right) / \partial P_j \right\} P_j + \mathcal{O}(P^2). $$

The first term in this series is the pose ${}^{ee}_{bs}T_m(q_i)$ derived from the model. Taking only the first and second terms into account yields an overdetermined set of linear equations in the $P_j$. These $P_j$ can then be fitted to the collected data in a "least-squares" sense. This means that the "distance" between the collected poses and the predictions made by the corrected model is minimal. Recall that no unique distance function for poses exists. In principle, this fact would not influence the calibration result, since one tries to make the distance zero, and a zero distance is defined unambiguously for any non-degenerate distance function. However, the distance is never exactly zero, due to measurement noise and/or an incomplete error parameter set.

Step 4 (Model correction). With the error estimates obtained in the previous step, one adapts the geometric model of the robot.

The calibration procedure requires (i) the robot to be taken off line for a significant period of time, and (ii) expensive external measurement devices. However, once calibrated, the adapted geometric model does not vary much anymore over time. In practice, robot calibration often yields good absolute accuracy, but in a limited subset of the robot's workspace only. Due to the presence of the error parameters, a calibrated robot always has a general kinematic structure, even if its nominal model is of the 321 type. Hence, calibrated robots definitely need the numerical kinematic procedures described later in this Chapter.

## Forward Velocity Kinematics

The forward velocity kinematics (FVK) solves the following problem:

Given the vectors of $(i)$ joint positions $q = (q_1 \cdots q_n)^T$ and (ii) joint velocities $\dot{q} = (\dot{q}_1 \cdots \dot{q}_n)^T$, what is the resulting end-effector twist $t^{ee}$?

The solution is always unique: one given set of joint positions and joint velocities always corresponds to only one single end-effector twist.

## The Jacobian Matrix

The relation between joint positions $q$ and end-effector pose T is nonlinear, but the relationship between the joint velocities $\dot{q}$ and the end-effector twist tee is linear : if one drives a joint twice as fast, the end-effector will move twice as fast too. Hence, the linear relationship can be represented by a matrix:

$$\underset{6\times1}{_{bs}t^{ee}} = \underset{6\times n}{_{bs}J(q)} \; \underset{n\times1}{\dot{q}}$$

The matrix $_{bs}J(q)$ is called the Jacobian matrix, or Jacobian for short, with respect to the reference frame {bs}. It was introduced by Withney. The terminology is in accordance with the "Jacobian matrix" as defined in classical mathematical analysis, (i.e., the matrix of partial derivatives of a function ) named after the Prussian mathematician Karl Gustav Jacob Jacobi. Note that the matrix of the linear mapping depends itself nonlinearly on the joint positions $q$. One often omits the explicit mention of $J$'s dependence on the joint positions $q$. The mapping from joint velocities to end-effector motion is unique, but different Jacobian matrices (i.e., coordinate representations) exist, depending on:

- A number of arbitrary choiced for the coordinates of the twist on the left-hand side of Eq.

  $\underset{6\times1}{_{bs}t^{ee}} = \underset{6\times n}{_{bs}J(q)} \; \underset{n\times1}{\dot{q}}$ : angular velocity on top, or linear velocity on top; velocity reference point in the base frame or in the end-effector frame; physical units of angular and linear velocity components; etc.

- The reference frame {bs} with respect to which the end-effector twist tee is expressed.

The physical interpretation of the Jacobian matrix follows immediately from a closer inspection of

Eq. $\underset{6\times1}{_{bs}t^{ee}} = \underset{6\times n}{_{bs}J(q)} \; \underset{n\times1}{\dot{q}}$ : the ith column of the Jacobian matrix is the end-effector twist generated by a unit velocity applied at the ith joint, and zero velocities at the other joints.

The Jacobian matrix is a basis for the vector space of all possible end-effector twists; hence, each column of the Jacobian is sometimes called a partial twist.

The twist interpretation of the Jacobian implies that the joint rates q˙ in Eq. $\underset{6\times1}{_{bs}t^{ee}} = \underset{6\times n}{_{bs}J(q)} \; \underset{n\times1}{\dot{q}}$ are dimensionless coordinates: they represent the magnitude of the twist with respect to the (arbitrarily defined) "unit twist."

Analytical Jacobian The matrix J in Eq. $\underset{6\times1}{_{bs}t^{ee}} = \underset{6\times n}{_{bs}J(q)} \; \underset{n\times1}{\dot{q}}$ is not a real mathematical Jacobian, since the angular velocity three-vector ω is not the time derivative of any three-vector orientation representation. Nevertheless, the time derivative of a forward position kinematic function $t_d = f(q)$ is well-defined:

$$\frac{dt_d}{dt} = \sum_{i=1}^{n} \frac{\partial f(q)}{\partial q_i} \frac{\partial q_i}{\partial t} \bar{J}\dot{q}.$$

The angular coordinates of the finite displacement twist td are a set of three Euler angles. The time derivatives of these Euler angles are related to the angular velocity three-vector by means of integrating factors. Hence, the difference between the Jacobian in Eq. $\underset{6\times1}{_{bs}t^{ee}} = \underset{6\times n}{_{bs}J(q)} \; \underset{n\times1}{\dot{q}}$ and the matrix of partial derivatives $\bar{J} = \partial f(q)/\partial q_i$ in Eq. $\frac{dt_d}{dt} = \sum_{i=1}^{n} \frac{\partial f(q)}{\partial q_i} \frac{\partial q_i}{\partial t} \bar{J}\dot{q}.$ are these integrating factors. $\bar{J}$

in Eq. (above) is sometimes called the analytical Jacobian, when it is necessary to distinguish it from the twist Jacobian J in Eq. $_{bs}t^{ee}_{6\times1} = {}_{bs}J(q)_{6\times n}\ \dot{q}_{n\times1}$.

## General FVK: Velocity Recursion

The previous Section defines the Jacobian matrix; this Section explains how to calculate it, starting from the knowledge about the current joint positions. The following procedure works for any serial structure with an arbitrary number of n joints. The basic idea is to perform an outward recursion (or "sweep"): one starts with the twist generated by the joint closest to the base, then transforms this twist to the second joint, adds the twist generated by this joint, transforms it to the third joint, etc.

## Numerical FVK

Step 0 Initialization. The twist of the "zeroth" joint in the base reference frame {0} is always zero:

$$i = 0, \text{ and } {}_0t^0 = \begin{pmatrix} \boldsymbol{0} \\ \boldsymbol{0} \end{pmatrix}.$$

Step 1 Recursion $i \rightarrow i+1$, until $i = n$:

Step 1.1 Transformation of the twist $_it^i$ to the next joint:

$$_{i+1}t^i = {}_{i+1}^i S\ _it^i,$$

where the screw transformation matrix $_{i+1}^i S$ is constructed from the (known) link transform $_{i+1}^i T$ (as described in another Chapter).

Step 1.2 Addition of the contribution of joint $i + 1$:

$$_{i+1}t^{\ i+1} = {}_{i+1}t^i + {}_{i+1}J_{i+1}\ \dot{q}_{i+1}.$$

The Jacobian column $_{i+1}J_{i+1}$ equals $(0\ 0\ 1\ 0\ 0\ 0)^T$ for a revolute joint, and $(0\ 0\ 1\ 0\ 0\ 0)^T$ for a prismatic joint, since the local $(0\ 0\ 0\ 0\ 0\ 0)^T$ Z-axis is defined to lie along the joint axis.

The result of the recursion is $_{n+1}t^{n+1} = {}_{ee}t^{ee}$, the total end-effector twist expressed in the end-effector frame {ee}.

Step 2 Transformation to the world frame {w} gives:

$$_\omega t^{ee} = {}_\omega^{ee}S\ _{ee}t^{ee}$$

The recursive procedure above also finds the Jacobian matrix: the second term in each recursion through Step 1.2 yields, for $\dot{q}_{i+1} = 1$, a new column of the Jacobian matrix, expressed in the local joint reference frame. Applying all subsequent frame transformations to this new Jacobian column results in its representation with respect to the world reference frame:

$$_\omega J_i = {}_1^\omega S\ _0^1 S\ _1^2 S \ldots {}_{i-1}^i S\ _iJ_i.$$

Variations on this FVK algorithm have appeared in the literature, differing only in implementation details to make the execution of the algorithm more efficient.

## Closed-form FVK for 321 Structure

For the 321 kinematic structure, more efficient closed-form solutions exist. The approach of the last reference is especially instructive, since it maximally exploits geometric insight. The wrist centre frame {4} of the 321 kinematic structure is the best choice as world frame, because it allows to solve the FVK by inspection, as the next paragraphs will show. (The Jacobian expressed in the wrist centre frame is sometimes called the "midframe" Jacobian.)

## Closed-form FVK

Step 1 The wrist is of the $ZXZ$ type, so the angular velocity generated by the fourth joint lies along the $Z^4$-axis. The angular velocity generated by the fifth joint lies along the $X^5$-axis, that is found by rotating the $X^4$-axis about $Z^4$ over an angle $q_4$. And the angular velocity generated by the sixth joint lies along the Z 6 -axis, whose orientation with respect to is found in the last column of Eq.

$$
{}^6_4T = \begin{pmatrix} {}^6_4R & 0_{3\times1} \\ 0_{1\times3} & 1 \end{pmatrix} = \begin{pmatrix} c_6c_4 - s_6c_5s_4 & -s_6c_4 - c_6c_5s_4 & -s_5s_4 & 0 \\ c_6s_4 + s_6c_5c_4 & -s_6s_4 + c_6c_5c_4 & s_5c_4 & 0 \\ -s_6s_5 & -c_6s_5 & c_5 & 0 \\ 0 & 0 & 0 & 1 \end{pmatrix}.
$$
In total, this yields

$$
{}_4J_{456} = \begin{pmatrix} {}_4J_4 & {}_4J_5 & {}_4J_6 \end{pmatrix} = \begin{pmatrix} 0 & c_4 & -s_5s_4 \\ 0 & s_4 & s_5c_4 \\ 1 & 0 & c_5 \\ 0 & 0 & 0 \\ 0 & 0 & 0 \\ 0 & 0 & 0 \end{pmatrix}.
$$

Step 2 The twists generated by joints 1, 2 and 3 are pure rotations too, but they cause translational velocities at the wrist centre point due to the non-zero lever arms between the joints and the wrist centre point. These moments arms are ${}_4p^{i,4}$ for $i=1,\ 2,3$, i.e., the position vectors from the three joints to the wrist centre point. Hence, inspection of Figure yields

$$
{}_4J_{123} = \begin{pmatrix} {}_4J_1 & {}_4J_2 & {}_4J_3 \end{pmatrix} = \begin{pmatrix} 0 & 1 & 1 \\ -s_{23} & 0 & 0 \\ c_{23} & 0 & 0 \\ -d^h & 0 & 0 \\ 0 & l_{2c_3} + l_3 & l_3 \\ 0 & l_{2s_3} & 0 \end{pmatrix}.
$$

Step 3 In order to obtain twists with the base frame as origin, it suffices to pre-multiply ${}_4J = \begin{pmatrix} {}_4J_{123} & {}_4J_{456} \end{pmatrix}$ by the screw transformation matrix ${}_{bs}S$:

$$
{}_{bs}J = {}_{bs}^4S \ {}_4J.
$$

$_{bs}^{4}S$ is straightforwardly derived from the solution to the forward position kinematics of the robot.

The motivation for choosing the wrist centre frame as reference frame is illustrated by the fact that the Jacobian expressed in this frame has a zero $3 \times 3$ submatrix.

Later, sections will need the value of the determinant of the Jacobian matrix. Also here, the advantage of the midframe Jacobian $_4J$ appears: it has a zero $3 \times 3$ submatrix, which enormously simplifies the calculation of the determinant:

$$\det(_4J) = \det \begin{pmatrix} -d^h & 0 & 0 \\ 0 & l_{2c3} + l_3 & l_3 \\ 0 & l_{2s3} & 0 \end{pmatrix} \det \begin{pmatrix} 0 & c_4 & -s_5 s_4 \\ 0 & s_4 & s_5 c_4 \\ 1 & 0 & c_5 \end{pmatrix}$$

$$= -d^h \, l_2 l_3 s_3 s_5.$$

Note that the determinant of the Jacobian is independent of the reference frame with respect to which it is calculated:

$$_fJ = {_f^iS} \, _iJ \implies \det(_fJ) = \det({_f^iS}) \det(_iJ),$$

and $\det(S) = \det^2(R) = 1$.

## Inverse Position Kinematics

The inverse position kinematics ("IPK") solves the following problem:

Given the actual end-effector pose $_{bs}^{ee}T$ , what are the corresponding joint positions $q = (q_1 \ldots q_n)^T$ ?

In contrast to the forward problem, the solution of the inverse problem is not always unique: the same end-effector pose can be reached in several configurations, corresponding to distinct joint position vectors. It can be proven, that a 6R manipulator (a serial chain with six revolute joints, as in Figs), with a completely general geometric structure, has sixteen different inverse kinematics solutions, found as the solutions of a sixteenth order polynomial.

As for the forward position and velocity kinematics, this Section presents both a numerical procedure for general serial structures, and the dedicated closed-form solution for robots of the 321 type, as described . Some older references describe similar solution approaches but in less detail.

The IK of a serial arm is more complex than its FK. However, many industrial applications don't need IK algorithms, since the desired positions and orientations of their end-effectors are manually taught: a human operator steers the robot to its desired pose, by means of control signals to each individual actuator; the operator stores the sequence of corresponding joint positions into the robot's memory; during subsequent task execution, the robot controller moves the robot to this set of taught joint coordinates. However, the current trends towards off-line programming does require IK algorithms, and hence calibrated robots. Recall that such calibrated robots have a general kinematic structure.

## General IPK: Newton-Raphson Iteration

Inverse position kinematics for serial robot arms with a completely general kinematic structure (but with six joints) are solved by iterative procedures, based on the Newton-Raphson approach:

## Numerical IPK

Step 1 Start with an estimate $\hat{q} = (\hat{q}_1 ... \hat{q}_6)^T$ of the vector of six joint positions. This estimate is, for example, the solution corresponding to a previous nearby pose, or, for calibrated robots, the solution calculated by the nominal model (using the procedure of the next Section if this nominal model has a 321 structure). As with all iterative algorithms, the better the initial guess, the faster the convergence.

Step 2 Denote the end-effector pose that corresponds to this estimated vector of joint positions $\hat{T}(\hat{q})$. The difference between the desired end-effector pose T (q) (with q the real joint positions which have to be found) and the estimated pose is "infinitesimal," as assumed in any iterative procedure:

$$T(q) = T(\hat{q}) \ T_\Delta(\Delta q).$$

$\Delta q = q - \hat{q}$ is the joint position increment to be solved by the iteration. Solving for $T_\Delta(\Delta q)$ yields

$$T_\Delta(\Delta q) = T^{-1}(\hat{q}) T(q).$$

Step 3 The coordinate expression of the infinitesimal pose $T_\Delta(\Delta q)$ is:

$$T_\Delta = \begin{pmatrix} 1 & -\delta_z & \delta_y & d_x \\ \delta_z & 1 & -\delta_x & d_y \\ -\delta_y & \delta_x & 1 & d_z \\ 0 & 0 & 0 & 1 \end{pmatrix}.$$

The infinitesimal displacement twist $t_\Delta(\hat{q}) = (\delta_x \ \delta_y \ \delta_z \ d_x \ d_y \ d_z)^T$ corresponding to $T_\Delta(\Delta q)$) is easily identified from Eq. $T_\Delta = \begin{pmatrix} 1 & -\delta_z & \delta_y & d_x \\ \delta_z & 1 & -\delta_x & d_y \\ -\delta_y & \delta_x & 1 & d_z \\ 0 & 0 & 0 & 1 \end{pmatrix}$. On the other hand, it depends linearly on the

joint increment $\Delta q$ through the Jacobian matrix $J(\hat{q})$, Eq. $\boxed{\underset{6\times1}{_{bs}t^{ee}} = \underset{6\times n}{_{bs}J(q)} \ \underset{n\times1}{\dot{q}}}$:

$$t_\Delta(\hat{q}) = J(\hat{q})\Delta q + \mathcal{O}(\Delta q^2).$$

$J(\hat{q})$ is calculated by the numerical FVK algorithm.

Step 4 Hence, the joint increment $\Delta q$ is approximated by

$$\boxed{\Delta q = J^{-1}(\hat{q}) t_\Delta(\hat{q}).}$$

The inverse of the Jacobian matrix exists only when the robot arm has six independent joints. Section 16 explains how to cope with the case of more or less than six joints.

Step 5 If $\Delta q$ is "small enough," the iteration stops, otherwise Steps 2–4 are repeated with the new estimate $\hat{q}_{i+1} = \hat{q}_i + \Delta q$.

This procedure gives an idea of the approach, but real implementations must take care of several numerical details, such as, for example:

1. Inverting a $6 \times 6$ Jacobian matrix (which is required in motion control) is not an insurmountable task for modern microprocessors (even if the motion controller runs at a frequency of 1000Hz or more), but nevertheless the implementation should be done very carefully, in order not to loose numerical accuracy.

2. In order to solve a set of linear equations $Ax = b$, it is, from a numerical point of view, not a good idea to first calculate the inverse $A^{-1}$ of the matrix $A$ explicitly, and then to solve the equation by multiplying the vector $b$ by this inverse, as might be suggested by Eq. $\Delta q = J^{-1}(\hat{q}) t_\Delta (\hat{q})$. Numerically more efficient and stable algorithms exist, the simplest being the Gaussian elimination technique and its extensions.

3. The numerical procedure finds only one solution, i.e., the one to which the iteration converges. Some more elaborate numerical techniques exist to find all solutions, such as for example the continuation, dialytic elimination and homotopy methods.

## Closed-form IPK for 321 Structure

The efficient closed-form IPK solution for the 321 structure relies again on the decoupling at the wrist centre point:

## Closed-form IPK

Step 1 The position of the wrist centre point is simply given by the inverse of Eq.
$$\boxed{{}_{bs}\boldsymbol{p}^{ee} = {}_{bs}\boldsymbol{p}^{wr} + {}_{bs}^{ee}\boldsymbol{R}\left(00l_6\right)^T.}:$$

$${}_{bs}\boldsymbol{p}^{wr} = {}_{bs}\boldsymbol{p}^{ee} - {}_{bs}^{ee}\boldsymbol{R}\left(00\ l_6\right)^T.$$

Step 2 Hence, the first joint angle is

$$q_1 = \operatorname{atan} 2\left({}_{bs}\boldsymbol{p}_x^{wr}, \pm {}_{bs}\boldsymbol{p}_y^{wr}\right).$$

The robot configuration corresponding to a positive ${}_{bs}\boldsymbol{p}_y^{wr}$ is called the "forward" solution, since the wrist centre point is then in front of the "body" of the robot; if ${}_{bs}\boldsymbol{p}_y^{wr}$ is negative, the configuration is called "backward."

Step 3 The horizontal and vertical distances dh and dv of the wrist centre point with respect to the shoulder frame {1} are found by inspection of figuer:

$$d^h = \sqrt{\left({}_{bs}\boldsymbol{p}_x^{wr}\right)^2 + \left({}_{bs}\boldsymbol{p}_y^{wr}\right)^2}, \ d^v = {}_{bs}\boldsymbol{p}_z^{wr} - l_1.$$

Step 4 Now, look at the planar triangles formed by the second and third links.

Step 4.1 The cosine rule gives:

$$q_3 = \pm \ \arccos\left(\frac{\left(d^h\right)^2 + \left(d^v\right)^2 - \left(l_2\right)^2 - \left(l_3\right)^2}{2l_2 l_3}\right).$$

A positive $q_3$ gives the "elbow up" configuration; the configuration with negative $q_3$ is called "elbow down."

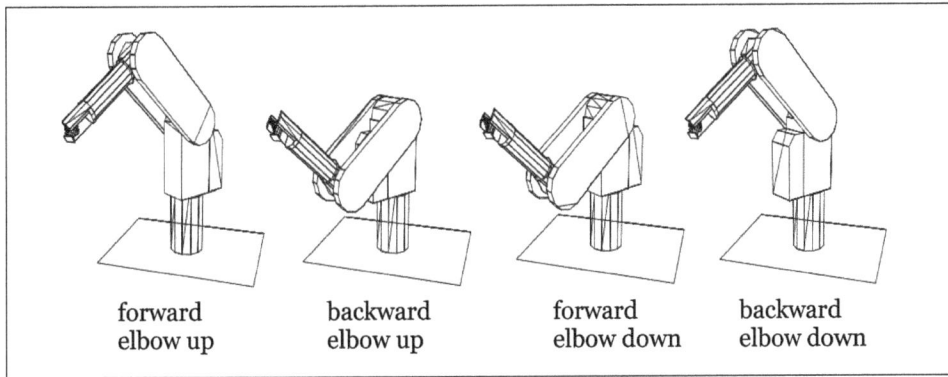

forward          backward          forward          backward
elbow up          elbow up          elbow down          elbow down

Four of the eight configurations corresponding to the same end effector pose,
for a 321 type of manipulator. The four other configurations are similar to these four,
except for a change in the wrist configuration from "flip" to "no flip."

Step 4.2 The tangent rules yield:

$$q_2 = \operatorname{atan} 2\left(d^h, d^v\right) - \alpha,$$

With

$$\alpha = \operatorname{atan} 2\left(l_3 s_3, \ l_2 + l_3 c_3\right)$$

Step 5 The inverse position for the $ZXZ$ wrist uses $_4^6 R$ as input, which is straightforwardly derived from $_{bs}^7 R$ (a known input parameter) and $_{bs}^4 R$ (which can be calculated as soon as the first three joint angles are known):

$$_4^6 R = _4^7 R = _4^{bs} R \ _{bs}^7 R,$$

With

$$_{bs}^4 R = R\left(Z, q_1\right) R\left(X, -q_2 - q_3\right).$$

Two solutions exist: one with $q_5 > 0$ (called the "no-flip" configuration) and one with $q_5 < 0$ (called the "flip" configuration).

In the algorithm above, a "configuration" corresponds to a particular choice of IPK solution. In total, the 321 manipulator has eight different configurations, by combining the binary decisions "forward/backward," "elbow up/elbow down," and "flip/no flip." Note that these names are not standardised: the robotics literature contains many alternatives.

## Inverse Velocity Kinematic

Assuming that the inverse position kinematics problem has been solved for the current end-effector pose $_{bs}^{ee}T$, the inverse velocity kinematics ("IVK") then solves the following problem:

Given the end-effector twist $t^{ee}$, what is the corresponding vector of joint velocities $\dot{q} = (\dot{q}_1 \ldots \dot{q}_n)^T$ ?

An alternative name for the IVK algorithm is the "resolved rate" procedure, especially in the context of robot control.

As in the previous Sections, a numerical procedure for general serial structures is given, as well as a dedicated closed-form solution for robots of the 321 type. The IVK problem is only well-defined if the robot has six joints: if n < 6 not all end-effector twists can be generated by the robot; if n > 6 all end-effector twists can be generated in infinitely many ways.

## General IVK: Numerical Inverse Jacobian

As for the inverse position kinematics, the inverse velocity kinematics for general kinematic structures must be solved in a numerical way. The simplest procedure corresponds to one iteration step of the numerical procedure used for the inverse position kinematics problem:

## Numerical IVK

Step 1 Calculate the Jacobian matrix $J(q)$.

Step 2 Calculate its inverse $J^{-1}(q)$ numerically.

Step 3 The joint velocities $\dot{q}$ corresponding to the end-effector twist $t^{ee}$ are:

$$\boxed{\dot{q} = J^{-1}(q)\, t^{ee}}$$

As mentioned before, better and more efficient algorithms calculate $\dot{q}$ without the explicit calculation of the matrix inverse $J^{-1}$.

## Closed-form IVK for 321 Structure

The symbolically derived Jacobian for the 321 kinematic structure turns out to be also easily invertible symbolically, when expressed in the wrist centre frame:

Step 1 The Jacobian $_4J$ has a zero 3×3 block in the lower right-hand side:

$$_4J = \begin{pmatrix} A & B \\ C & 0_3 \end{pmatrix}$$

**Step 2** It is then easily checked by straightforward calculation that:

$$_4J^{-1} = \begin{pmatrix} 0_3 & C^{-1} \\ B^{-1} & -B^{-1}AC^{-1} \end{pmatrix}.$$

**Step 3** The inverses $B-1$ and $C-1$ are found symbolically by dividing the transpose of their matrices of cofactors by their determinants. These determinants are readily obtained from Eqs

$$_4J_{456} = \begin{pmatrix} _4J_4 & _4J_5 & _4J_6 \end{pmatrix} = \begin{pmatrix} 0 & c_4 & -s_5s_4 \\ 0 & s_4 & s_5c_4 \\ 1 & 0 & c_5 \\ 0 & 0 & 0 \\ 0 & 0 & 0 \\ 0 & 0 & 0 \end{pmatrix}. \quad _4J_{123} = \begin{pmatrix} _4J_1 & _4J_2 & _4J_3 \end{pmatrix} = \begin{pmatrix} 0 & 1 & 1 \\ -s_{23} & 0 & 0 \\ c_{23} & 0 & 0 \\ -d^h & 0 & 0 \\ 0 & l_{2c_3} + l_3 & l_3 \\ 0 & l_{2s_3} & 0 \end{pmatrix}.$$

$$\det(B) = s_5, \quad \det(C) = l_2 l_3 d^h \, s_3.$$

Hence,

$$B^{-1} = \frac{1}{s_5} \begin{pmatrix} s_4c_5 & s_5c_4 & -s_4 \\ -c_5c_4 & s_5s_4 & c_4 \\ s_5 & 0 & 0 \end{pmatrix}, \quad C^{-1} = \begin{pmatrix} \dfrac{-1}{dh} & 0 & 0 \\ 0 & 0 & \dfrac{1}{l_2 s_3} \\ 0 & -\dfrac{1}{l_3} & -\dfrac{l_2c_3 + l_3}{l_2 l_3 s_3} \end{pmatrix}.$$

**Step 4** In order to find the joint velocities, one has to post-multiply $_4J^{-1}$ by $_4^{bs}S$:

$$\dot{q} = _4J^{-1} \, _4^{bs}S \, _{bs}t = _{bs}J^{-1} \, _{bs}t.$$

Assuming that the inverse position kinematics problem has been solved for the current end-effector pose $_{bs}^{ee}T$, the inverse force kinematics ("IFK") then solves the following problem:

Given the end-effector twist $\omega^{ee}$ that acts on the end-effector, what is the vector of joint forces/torques $\tau = (\tau_1 \cdots \tau_n)^T$ that keeps $\omega^{ee}$ in static equilibrium?

## Projection on Joint Axes

The end-effector twist $t^{ee}$ is the sum of the twists generated by all joints individually; but a wrench w exerted on the end-effector is transmitted unchanged to each joint in the serial chain. Part of the transmitted wrench is to be taken up actively by the joint actuator, the rest is taken up passively by the mechanical structure of the joint. While the wrench is physically the same screw at each joint, its coordinates expressed in the local joint frames differ from frame to frame. The wrench coordinates $_{bs}\omega$ of the end-effector wrench expressed in the base reference frame {bs} are related to the coordinates $_i\omega$ of the wrench expressed in the reference frame of the $i$ ith joint through the screw transformation matrix $_i^{bs}S$:

$$_i\omega = {_i^{bs}}S \; _{bs}\omega.$$

If this local frame {i} has its $Z^i$ axis along the prismatic or revolute joint axis, then the force component $\tau_i$ felt by the joint actuator corresponds to, respectively, the third and sixth coordinate of $_i\omega$. These coordinates are found by premultiplying $_{bs}\omega$ by the third or sixth rows of $_i^{bs}S$, or equivalently, the third and sixth columns of $_i^{bs}S^T$:

$$_i^{bs}S_{3\times} = {_i^{bs}}S_{\times3}^T = \begin{pmatrix} _{bs}e_z^i \\ 0 \end{pmatrix}, \text{ and } {_i^{bs}}S_{6\times} = {_i^{bs}}S_{\times6}^T = \begin{pmatrix} _{bs}p^{bs,i} \times {_{bs}}e_z^i \\ _{bs}e_z^i \end{pmatrix},$$

where $S_{3\times}$ indicates the third row of matrix $S$, and $S_{\times6}$ the sixth column. These columns resemble the columns $J_i$ of the Jacobian matrix as used for screw twists, but with the first and second three-vectors interchanged. So, premultiplication of $J_i$ by

$$\tilde{\Delta} = \begin{pmatrix} 0_{3\times3} & 1_{3\times3} \\ 1_{3\times3} & 0_{3\times3} \end{pmatrix}$$

makes the resemblance exact. The above reasoning can be repeated for all joints, and for all other twist and wrench representations. Hence, the IFK is

$$\boxed{\tau = \left(\tilde{\Delta}\,J\right)^T \omega.}$$

The $\tilde{\Delta}$ is solely an artifact coming from the choice of mathematical representation, and it disappears if one chooses to represent twists (or wrenches) by six-dimensional coordinate vectors that have their linear and angular components in the other place with respect to the convention followed in this text. Equation $\boxed{\tau = \left(\tilde{\Delta}\,J\right)^T \omega.}$ is often referred to as the "Jacobian transpose" relationship between end-effector wrench w and joint force/torque vector $\tau$. It represents the fact that the joint torque that keeps a static wrench exerted on the end-effector in equilibrium is given by the projection of this end-effector wrench on the joint axis. This fact is valid for any serial robot arm.

In the robotics literature you see this relationship most often in the form $\tau = J^T\omega$, due to the difference in twist representation from the one used in this text: $J^{\text{literature}} = \tilde{\Delta}J^{\text{this text}}$.

Strictly speaking, $\tilde{\Delta}J$ is not a "Jacobian matrix," since wrenches are not the partial derivatives of anything.

## Conservation of Virtual Work

A second approach to derive this IFK is through the instantaneous power generated by an end-effector twist t against the wrench w exerted on the end-effector. This power equals $t^T\tilde{\Delta}\omega$ in Cartesian space coordinates, and $\sum_{i=1}^{n}\tau_i\dot{q}_i$ in joint space coordinates. Replacing t by $J\dot{q}$. and some simple algebraic manipulations yield Eq. $\tau = \left(\tilde{\Delta}\,J\right)^T \omega.$ again.

## Dual Twist–Wrench Bases

For a robot with six joints, the Jacobian matrix is a basis for the twist spac of the end-effector. A dual basis for the wrench space exists when a basis for the twist space is given. The kinematic structure of the robot is a de facto choice of twist space basis. The natural pairing ("reciprocity") between twists and wrenches leads to a de facto dual basis in the wrench space, whose physical interpretation is as follows: the ith column of the "dual" wrench basis of a serial robot arm is the wrench on the end-effector that generates a unit force/torque at the ith joint, and zero forces/torques at the other joints. Each column of the dual wrench basis is sometimes called a partial wrench.

This text uses the notation $G = (G_1 \ldots G_6)$ for the matrix of the six dual basis wrenches $G_i$. Its formal definition follows from the following relationship with the Jacobian matrix J of the same robot arm:

$$J \tilde{\Delta} \, G = 1_{6 \times 6}.$$

Since $G$ is a basis for the wrench space, each wrench $\omega$ on the end-effector has coordinates $(\tau_1, \ldots, \tau_6)$:

$$\omega = G\tau.$$

$\tau_1$ is the force/torque required at the ith joint to keep the end-effector wrench w in static equilibrium (neglecting gravity of the links!). The relation with the "Jacobian transpose" formula for the Inverse Force Kinematics, Eq. $\tau = \left( \tilde{\Delta} \, J \right)^T \omega$, is also immediately clear:

$$G = \left( \tilde{\Delta} J \right)^{-T}.$$

## Forward Force Kinematics

The forward force kinematics ("FFK") solves the following problem:

Given the vectors of joint force/torques $\tau = (\tau_1 \ldots \tau_n)^T$, what is the resulting static wrench $\omega^{ee}$ that the end-effector exerts on the environment?

This problem is only well-defined if the robot has six joints, and if the end-effector is rigidly fixed to a rigid environment! If n < 6 the robot cannot generate a full six-dimensional space of end-effector wrenches; if n > 6 all wrenches can be generated in infinitely many ways. So, most often, the FFK is a dubious "property" of the robot: it depends equally much on the dynamic properties of the environment with which the robot is in contact.

## Closed-form FFK for 321 Structure

As before, using the wrist centre point of the 321 kinematic structure allows for a solution of the FFK problem by simple inspection, since the partial wrench of each joint is easily found from.

Joint 1 The partial wrench is a pure force through the wrist centre point and parallel to the axes of the second and third joints.

Joint 2 The partial wrench is a pure force through the wrist centre point and through the joint axis of the first and third joints.

Joint 3 The partial wrench is a pure force through the wrist centre point and through the joint axis of the first and second joints.

Joints 4,5,6 The partial wrench of each of these joints is the combination of:

1. A pure moment about a line through the wrist centre point and orthogonal to the axes of the two other wrist joints. This moment has no components about these other two joint axes. However, it can have components about the first three joint axes.

2. These components about the first three joint axes are compensated by pure forces that do not generate moments about these first three joint axes. These forces are: (i) through the origins of the second and third joint frames (i.e. along $l_2$), and (ii) through the first joint axis and parallel with the second and third joint axes.

## Singularities

The inverse velocity kinematics exhibit singularities, whose physical interpretation is that, at such a singularity, the Jacobian matrix $J$ looses rank.

This means that the end-effector looses one or more degrees of twist freedom (i.e., instantaneously, the end-effector cannot move in these directions). Equivalently, the space of wrenches on the end-effector that are taken up passively by the mechanical structure of the robot (i.e., without needing any joint torques to be kept in static equilibrium) increases its dimension.

These singularities are physical, in the sense that they have a physical origin, and are found in any mathematical representation used to describe the kinematics. However, in addition to the physical singularities, representational singularities exist too, for example, the FVK or IVK algorithms exhibit one of the mathematical singularities inherent in any minimal coordinate representation.

The mathematical discussion of singularities relies on the rank of the Jacobian matrix $J$, which, for a serial manipulator with n joints, is a $6 \times n$ matrix. Two distinct cases occur:

Not full column rank Or, rank $(J) <$ n. This case is discussed in more detail in Section 16. The Jacobian $J$ has a non-empty null space, and any Cartesian velocity (in the range of $J$ !) can be generated by an infinite and continuous set of joint velocities. This redundancy singularity condition is, for example, always satisfied for manipulators with more than six joints.

Not full row rank Or, rank$(J) < 6$. This means that $J$ cannot span the full Cartesian velocity twist space anymore: one or more Cartesian twist velocity degrees of freedom disappear since a set of joint axes becomes linearly dependent. This fact is reflected in the inverse velocity algorithm by joint speeds that tend to infinity. The solution in these cases can be twofold: (i) change the motion a bit, in order to avoid the singularity, or (ii) reduce the speed of the robot to zero when it reaches the singularity.

This last kind of singularity is called "configuration singularity", or "singularity" for short. The word "singularity" very often also denotes the configuration in which the loss of Jacobian rank occurs. A robot in a singular configuration is a special type of constrained system. Serial robots with less than six independent joints are always "singular" in the sense that they can never span a six-dimensional twist space. This is often called an "architectural singularity". A singularity is usually not an isolated point in the workspace of the robot, but a sub-manifold with an infinite number of continuously connected singular positions.

If one or more degrees of motion freedom disappear in a singularity, a corresponding number of reciprocal screws appear. The reciprocal wrench space is at least one-dimensional, or increases in dimension if it was already non-empty outside of the singularity. The wrenches in the reciprocal wrench space are transmitted from the end-effector to the base completely passively, i.e., just by the mechanical structure of the robot without any need for joint forces/torques.

## Numerical Singularity Detection

As for the inverse position kinematics, the general approach to find the singularities of a serial manipulator is numerical. For 321 robots, however, a closed-form solution follows straightforwardly from the closed-form velocity kinematics.

Solution follows straightforwardly from the closed-form velocity kinematics. For a square Jacobian, $\det(J) = 0$ is a necessary and sufficient condition for a singularity to appear. However, some robots do not have square Jacobians. Hence, a better numerical criterion is required, i.e., one that finds the changes in the rank of the Jacobian. The most numerically stable criterion is based on the Singular Value Decomposition (SVD), that works for all possible kinematic structures: every matrix A (with arbitrary dimensions $m \times n$) has an SVD decomposition of the form:

$$\underset{m \times n}{A} = \underset{m \times m}{U} \ \underset{m \times n}{\Sigma} \ \underset{n \times n}{V} \ \ with \ \ \Sigma = \begin{pmatrix} \sigma_1 & 0 & \cdots & 0 & 0 & \cdots & 0 \\ 0 & \sigma_2 & \cdots & 0 & 0 & \cdots & 0 \\ \vdots & \vdots & \ddots & 0 & 0 & \cdots & 0 \\ 0 & 0 & \cdots & \sigma_m & 0 & \cdots & 0 \end{pmatrix},$$

represented here for $n > m$. U and V are orthogonal matrices, and the singular values $\sigma_i$ are in descending order: $\sigma_1 \geq \sigma_2 \geq \cdots \geq \sigma_m \geq 0$. A has full rank (i.e., rank $(A) = m$ in the case above where $n > m$) if $\sigma_m \neq 0$; it loses rank if $\sigma_m \approx 0$, i.e., $\sigma_m$ is zero within a numerical tolerance factor. Hence, the most popular way to monitor a robot's "closeness" to singularity is to check the smallest singular value in the SVD of its Jacobian matrix. This requires some computational overhead (of $O(n^3)$), and efficient methods exist for closed-form kinematics designs, such as the 321 structure.

## Singularity Detection for 321 Structure

The singular positions of the 321 robot structure, Fig. follow immediately from the closed-form inverse velocity kinematics: the determinant of the Jacobian is $-d^h l_2 l_3 s_3 s_5$, equation

$$\det(_4 J) = \det \begin{pmatrix} -d^h & 0 & 0 \\ 0 & l_{2c3} + l_3 & l_3 \\ 0 & l_{2s3} & 0 \end{pmatrix} \det \begin{pmatrix} 0 & c_4 & -s_3 s_4 \\ 0 & s_4 & s_5 c_4 \\ 1 & 0 & c_5 \end{pmatrix}.$$

$$= -d^h \ l_2 l_3 s_3 s_5.$$

Hence, it vanishes in the following three cases:

Arm-extended singularity $(q_3 = 0)$ (calls it a "regional" singularity.) The robot reaches the end of its regional workspace, i.e., the positions that the wrist centre point can reach by moving the first three joints. The screw reciprocal to the remaining five motion degrees of freedom is a force along the arm.

Wrist-extended singularity $(q_5 = 0)$ calls it a "boundary" singularity.) The first and last joint of the wrist are aligned, so they span the same motion freedom. Hence, the angular velocity about the common normal of the three wrist joints is lost. The screw reciprocal to the remaining five motion degrees of freedom cannot be described in general: it depends not only on the wrist joints, but on the first three joint angles too.

Wrist-above-shoulder singularity $(d^h = 0)$ ( calls it an "orientation" singularity.) The first joint axis intersects the wrist centre point. This means that the three wrist joints (which are equivalent to a spherical joint) and the first joint are not independent. The screw reciprocal to the five remaining motion degrees of freedom is a force through the wrist centre point, and orthogonal to the plane formed by the first three links.

Contrary to what might be suggested by the previous paragraphs, it is not necessary that a robot passes through a singularity in order to change configuration. Only special structures, such as the 321 robots, have their singularities coinciding with their configuration borders.

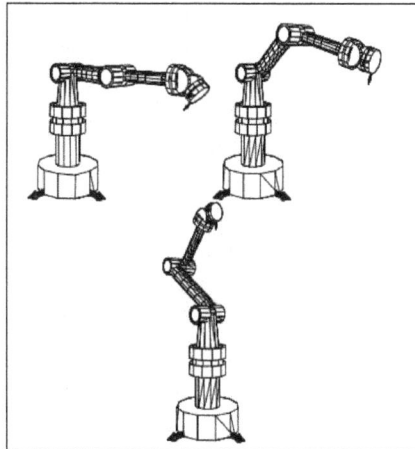

The three singular positions for the 6R wrist-partitioned serial robot arm with closed form kinematic solutions: "arm-extended," "wrist-extended" and "wrist-above-shoulder."

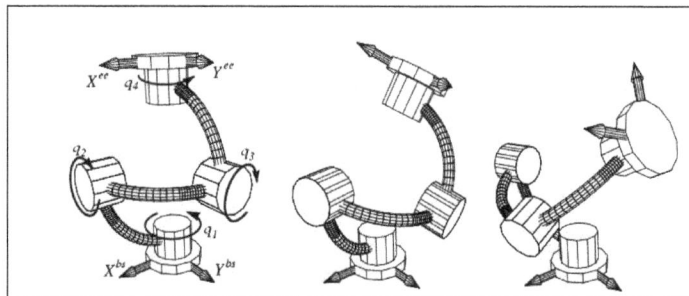

Redundant wrist with four revolute joints. In the left-most configuration, two axes line up, but the wrist does not become singular.

The zero block in the Jacobian matrix for a 321 design, equation. $_4J = \begin{pmatrix} A & B \\ C & 0_3 \end{pmatrix}$, comes from the fact that the wrist is spherical, i.e., it generates no translational components when expressed in the wrist centre frame. A spherical wrist does not only decouple the position and orientation kinematics, but also the singularities of the wrist, $\det(B) = 0$, and the singularities of the regional structure of the arm, $\det(C) = 0$.

A general kinematic structure has more complicated and less intuitive singularities. The reason why shoulder offsets have been introduced as extensions to the 321 kinematic structure is that they make sure that the robot cannot reach "wrist-above-shoulder" singular position. The reason behind elbow offsets is to avoid the "arm-extended" singularity in the "zero position" of the robot; zero positions (of part of the arm) are often used as reference positions at start-up of the robot, and it is obviously not a good idea to let the robot start in a singularity.

## Redundancy

A manipulator with n joints is called redundant if it is used to perform a task that requires less than the n available degrees of freedom. For example, a classical six degrees of freedom serial robot is redundant if it has to follow the surface of a workpiece with a vertex-like tool and no orientation constraints are specified: only three joints are required to keep the tool vertex on the surface, and only five are needed to keep the tool aligned with, for example, the normal direction to the surface. In general, however, robots are designed for more than just one single task, and hence we speak of redundant robots when they have seven or more joints. An obvious choice for an anthropomorphic redundant robot is the "7R" manipulator, Fig. It has an extra joint between the "shoulder" and the "elbow" of the 6R wrist-partitioned manipulators in Figures. In this way, the robot can reach "around" obstacles that the 6R robots cannot avoid. Such a redundant manipulator can attain any given pose in its dextrous workspace in infinitely many ways. This is obvious from the following argument: if one fixes one particular joint to an arbitrary joint value, a full six degrees of freedom robot still remains, and this robot can reach the given pose in at least one way. This 7R manipulator can also avoid the "extended-wrist" and "wrist-above-shoulder" singularities (but not necessarily both at the same time).

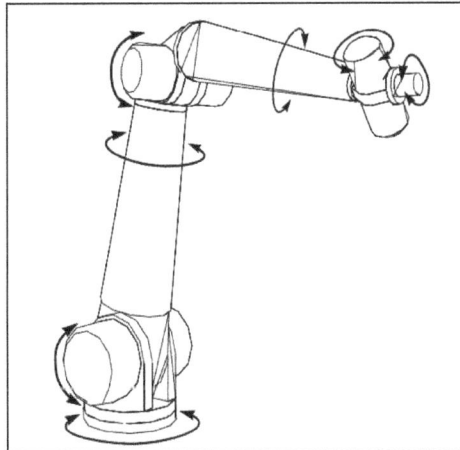

Redundant serial arm with seven revolute joints. It differs
from the 321 structure in having an extra shoulder joint.

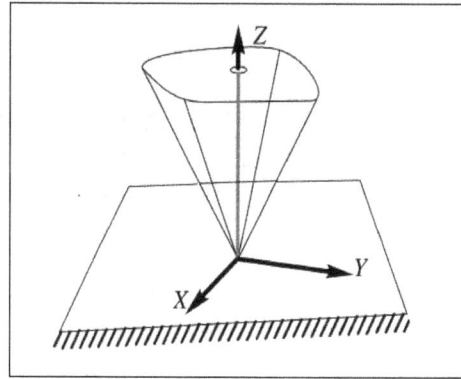

Vertex-face contact.

## Forward Kinematics

The forward kinematics (position, velocity, acceleration) for a redundant robot add no complexity to the case of non-redundant robots: any given set of joint positions and joint velocities still corresponds to one unique end-effector pose and twist, and the forward velocity kinematics are still described by the Jacobian matrix:

$$\underset{6\times1}{t^{ee}} = \underset{6\times7}{\boldsymbol{J}(\boldsymbol{q})}\ \underset{7\times1}{\dot{q}}\ .$$

This Jacobian matrix has more columns than rows, e.g., it is a $6 \times 7$ matrix in the case of the 7R robot. Hence, it always has a null space, i.e., a set of joint velocities that do not move the end-effector:

$$\boxed{\text{Null}\ \left(\boldsymbol{J}(\boldsymbol{q})\right)\ =\ \left\{\ \dot{q}^{N}\ |\ \boldsymbol{J}(\boldsymbol{q})\ \dot{q}^{N} = 0\ \right\}.}$$

This null space depends on the current joint positions. Equation (50) implies that an arbitrary vector of the null space of the Jacobian can be used as an internal motion of the robot:

$$t^{ee} = \boldsymbol{J}(\boldsymbol{q})\dot{q} = J(\boldsymbol{q})\left(\dot{q} + \dot{q}^{N}\right)\ .$$

In other words, the internal velocity joint vector $\dot{q}^{N}$ only moves the links of the robot, but not its endpoint.

## Inverse Kinematic

The inverse kinematics of a redundant robot require the user to specify a criterion with which to solve the ambiguities in the joint positions and velocities (internal motions) corresponding to the specified end-effector pose and twist. Some examples of redundancy resolution criterions are:

1. To keep the joints as close as possible to a specified position. The goal of this criterion is to avoid that joints reach their mechanical limits. A simple approach to reach this goal is to attach virtual springs to the joints, with the equilibrium position of the springs near the middle of the motion range of the joints. With this spring model, the redundancy resolution criterion corresponds to the minimization of the potential energy in the springs.

2. To minimize the kinetic energy of the manipulator. This approach requires the knowledge of the mass distribution of all links (and also of the motor shafts, in the case of electrical motors and high gear ratios).

3. To maximize the manipulability of the manipulator, i.e., keep the robot close to the joint positions that give it the best ability to move and/or exert forces in all directions. This requires some quantitative "measure" of motion ability; it is well-known that not no such natural measures exist, so this criterion involves arbitrary choices.

4. To minimize the joint torques required for the motion. The goal of this criterion is to avoid saturation of the actuators, and to execute the task with minimum "effort,". This criterion is very relevant for "bionic" robots, such as protheses, or robotic models for human motion.

5. To execute a high priority task while using the redundancy to achieve a lower priority task in parallel. In other words, the lower priority task can be executed in the null space of the high priority task.

6. To avoid obstacles in the robot's workspace. For example, a robot with an extra shoulder or elbow joint can reach "around" obstacles. This approach is similar to the criterion of keeping the joints away from their limits; the difference is that the (repelling) "springs" are now attached between the robot links and the environment objects to avoid.

7. To avoid singularities in the robot kinematics. For example, the 4R "Hamilton wrist," can avoid the "extended-wrist" singularity. The addition of an extra joint is by itself not sufficient to have singularity-free motion, in the sense that even a redundant robot can end up in a singularity. The "internal spring" approach mentioned before is one possible way to automatically generate singularity-free motions in this case, but it is in general difficult to find the "obvious" locations to attach the virtual spring to.

8. To travel through a singularity while keeping the joint velocities bounded.

Some of these redundancy resolution criterions can be solved via the concept of the extended Jacobian. This approach starts from the observation that the non-square $6 \times n$ Jacobian can be made into a square $n \times n$ matrix by extending it with $n - 6$ rows, collected in a $(n-6) \times n$ matrix $A$:

$$\bar{J} = \left( \frac{J(q)}{A(q)} \right).$$

This is equivalent to adding $n - 6$ linear constraints on the joint velocities:

$$A(q)\dot{q} = 0.$$

In order to obtain a full-rank extended Jacobian $\bar{J}$, the constraint matrix $A$ must be full rank by itself, and, in addition, be transversal (or "transient") to the Jacobian J, i.e., the null spaces of $A$ and $J$ should have no elements in common. Equation $\bar{J} = \left( \frac{J(q)}{A(q)} \right)$. then has a uniquely defined inverse:

$$\bar{J}^{-1} = (B \mid *).$$

Robotics: Design, Construction and Applications

The $*$ indicates the part of the matrix $\bar{J}^{-1}$ which is irrelevant to the discussion in the following paragraphs. The relevant $n \times 6$ matrix $B$ is a so-called generalized inverse, or pseudo-inverse, often denoted by $B = J^{\dagger}$,: it satisfies $JB = 1_{6 \times 6}$ and $BJ = 1_{n \times n}$. (This follows straightforwardly from the definition of $\bar{J}$.) With it, the forward velocity kinematics, Eq. $\underset{6 \times 1}{t^{ee}} = \underset{6 \times 7}{J(q)} \underset{7 \times 1}{\dot{q}}$ , can be "inverted":

$$\dot{q} = B t^{ee} .$$

Do not forget that the resulting joint velocities depend on the choice of the constraint matrix $A$. The following paragraphs derive this general result of Eq. $\dot{q} = B t^{ee}$ in more detail, and in an alternative way for the particular example of the kinetic energy minimization criterion. The kinetic energy T of a serial manipulator is of the form:

$$T = \frac{1}{2} \dot{q}^T M(q) \dot{q} .$$

The inertia matrix $M$ is both invertible and symmetric, because $T$ is a positive scalar and hence $T^T = T$. Minimizing the kinetic energy, while at the same time obeying the inverse kinematics requirement that $t^{ee} = J\dot{q}$ , transforms the solution to the following constrained optimization problem:

$$\begin{cases} \min_{\dot{q}} T = \frac{1}{2} \dot{q}^T M(q) \dot{q} \\ \text{such that } t^{ee} = J(q)\dot{q} \end{cases}$$

The classical solution of this kind of problem uses Lagrange multipliers, i.e., the constraint in (57) is integrated into the functional $T$ to be minimized as follows:

$$\min_{\dot{q}} T' = \frac{1}{2} \dot{q}^T M(\dot{q}) + \lambda^T \left( t^{ee} - J\dot{q} \right).$$

(For notational simplicity, we dropped the dependence of $M$ and $J$ on the joint positions q.) $\lambda$ is the column vector of the (currently unknown) Lagrange multipliers. They can be physically interpreted as the impulses (forces times mass) generated by violating the constraint $t^{ee} - J\dot{q} = 0$ (Check the physical units!) The Lagrange multipliers are determined together with the desired joint velocities by setting to zero the partial derivatives of the functional $T'$ with respect to the minimization parameter vector $\dot{q}$ :

$$\underset{1 \times 7}{\dot{q}^T} \underset{7 \times 7}{M} - \underset{1 \times 6}{\lambda^T} \underset{6 \times 7}{J} = \underset{1 \times 7}{0}$$

This gives a set of seven equations, in the seven joint velocities and the six Lagrange multipliers. These Lagrange multipliers can be solved for by postmultiplying Eq. $\underset{1 \times 7}{\dot{q}^T} \underset{7 \times 7}{M} - \underset{1 \times 6}{\lambda^T} \underset{6 \times 7}{J} = \underset{1 \times 7}{0}$ by $M^{-1}J^T$:

$$\dot{q}^T J^T = \lambda^T \left( J M^{-1} J^T \right).$$

The left-hand side of this equation equals the transpose of the end effector twist, $\left( t^{ee} \right)^T$, and the matrix triplet on the right-hand side is a square 6×6 matrix that can be inverted (at least if the manipulator is not in a singular configuration, Sect. 15). Hence,

$$\lambda^T = \left(t^{ee}\right)^T \left(\boldsymbol{J}\boldsymbol{M}^{-1}\boldsymbol{J}^T\right)^{-1}.$$

Equations $\underset{1\times7}{\dot{\boldsymbol{q}}^T}\ \underset{7\times7}{\boldsymbol{M}} - \underset{1\times6}{\lambda^T}\ \underset{6\times7}{\boldsymbol{J}} = 0_{1\times7}$ and $\lambda^T = \left(t^{ee}\right)^T \left(\boldsymbol{J}\boldsymbol{M}^{-1}\boldsymbol{J}^T\right)^{-1}$, and the fact that $\boldsymbol{M}$ is symmetric, yield

$$\dot{\boldsymbol{q}} = \boldsymbol{M}^{-1}\boldsymbol{J}^T \left(\boldsymbol{J}\boldsymbol{M}^{-1}\boldsymbol{J}^T\right)^{-1} t^{ee}$$

$$= \boldsymbol{J}^{\dagger}_{M-1} t^{ee}.$$

$\boldsymbol{J}^{\dagger}_{M-1}$ is a $n \times 6\ (n > 6))$ matrix, the so-called weighted pseudo-inverse of $\boldsymbol{J}$, with $\boldsymbol{M}^{-1}$ acting as weighting matrix on the space of joint velocities. It is not a good idea to calculate the solution q˙ by the straightforward matrix multiplications of Eq. $\dot{\boldsymbol{q}} = \boldsymbol{M}^{-1}\boldsymbol{J}^T \left(\boldsymbol{J}\boldsymbol{M}^{-1}\boldsymbol{J}^T\right)^{-1} t^{ee}$; better numerical techniques exist.

The redundancy resolution approaches based on the Jacobian matrix yield only local optimality. For example, one minimizes the instantaneous kinetic energy, not the kinetic energy over a complete motion. The success of the extended Jacobian approach is due to the fact that analytical solutions exist for quadratic cost functions only.

Cyclicity—Holonomic constraints. When one steers the end-effector of a redundant robot along a cyclic motion (i.e., it travels through the some trajectory of end-effector poses repetitively), the pseudo-inverse derived from an extended Jacobian typically results in different joint trajectories during each cycle. Whether or not the joint space trajectory is cyclic depends on the integrability of the constraint equation $A(q)\dot{q} = 0$. If this equation is integrable, the constraints are called holonomic. This name comes from the Greek word holos, which means "whole, integer." A (non) holonomic constraint on the joint velocities can (not) be integrated to give a constraint on the joint positions.

It can be proven that using the mass matrix of the robot as weighting in the weighted pseudo-inverse approach gives a cyclic result.

## Constrained Manipulator

The previous Section looked at the case in which one imposes virtual constraints on the manipulator; this Section explains how to deal with physical constraints: a free-space motion specified in 6D while only less than six joints are available, or motion with a six-jointed manipulator in contact with a stiff environment. The former gives rise to an optimization problem in Cartesian space (i.e., how to approximate the specified motion "as good as possible"; the latter gives rise to an optimization problem in joint space, dual to the redundancy resolution in the previous Section (i.e., how to use all six joints to "optimally" execute a motion in a less than six dimensional constraint space). We distinguish between:

1. Kinematic constraints (also called geometric constraints): the instantaneous velocities that the robot can execute form (part of) a vector space with dimension lower than six. This space is

called the twist space of the constraint. Equivalently, there exists a non-empty wrench space of generalized forces at the end-effector that are balanced passively by the mechanical structure of the robot. "Passive" means: without requiring torques at the driven joints. The twist and wrench spaces are always reciprocal. An element of the wrench space is said to be a reciprocal wrench.

2. Dynamic constraints: the actuators cannot produce sufficient torque to generate a desired velocity, or rather acceleration. This means that the bandwidth of the robot motion is limited, but not necessarily the spatial directions in which it can move.

A mechanical limit of a revolute joint is a simple example of a kinematic constraint: when the joint has reached this mechanical limit, the end-effector can resist any wrench that corresponds to a pure torque about this joint (and in the direction of the mechanical limit!). Another simple example of a kinematically constrained robot is a robot with less than six joints, e.g., the SCARA robot of Fig. The twist space of this robot is never more than four-dimensional: it can always resist pure moments about the end-effector's $X$ and $Y$ axes (if $Z$ is the direction of the translational and angular motion of the last link).

A kinematic motion constraint is correctly modelled by (i) a basis for the twist space of instantaneous velocities allowed by the constraint, or (ii) a basis of the wrench space of instantaneous forces the constraint can absorb. A kinematic constraint is not correctly modelled by a so-called "space of impossible motions" (i.e, motions that the robot cannot execute) or a "space of non-reciprocal screws" (i.e., wrenches that are not reciprocal to the constraint twist space): neither of these concepts is well-defined, since (i) the sum of an impossible motion with any possible motion remains an impossible motion, and (ii) adding any reciprocal screw to a non-reciprocal screw gives another non-reciprocal screw.

The discussion in this Section assumes that all bodies and all kinematic constraints are infinitely stiff and have no friction. If this is not the case, the robot can move against such a compliant environment and will feel force in the direction of motion; hence, the actual relationships between the possible motions and the corresponding forces are determined by the contact impedance between robot and environment.

## Free-space Motion with Less than Six Degrees of Freedom

Assume the manipulator has less than six joints, say $6 - n$. Hence, the Jacobian $J$ is a $6 \times n$ matrix, and there always exists a reciprocal wrench space of at least dimension n. Such a manipulator is constrained to move on a $(6 - n)$-dimensional sub-manifold of the six-dimensional space of translations and rotations. That means that it can not generate any arbitrary end-effector twist $t^{ee}$. A kinematic energy based pseudo-inverse procedure exists to project $t^{ee}$ on the span of $J$. This procedure is derived quite similarly to the redundancy resolution procedure of the previous Section; nevertheless, it has fundamentally different properties. The (unconstrained) objective kinetic energy function to be minimized is:

$$\min_{\dot{q}} T = \frac{1}{2}\left(t^{ee} - J\dot{q}\right)^T M\left(t^{ee} - J\dot{q}\right),$$

with $M$ a (full-rank) Cartesian space mass matrix. (The choice of this mass matrix is in most cases fully arbitrary!) The physical interpretation of this minimization problem is that the end-effector twist $t^{ee}$ is approximated by that twist $J\dot{q}$ on the constrained sub-manifold that results in the mallest "loss" of kinetic energy (of the virtual Cartesian mass) compared to the case that the full $t^{ee}$ could be executed. Setting the partial derivative of the objective function with respect to the joint velocities to zero yields:

$$Mt^{ee} = MJ\dot{q} .$$

(Recall that $M$ is symmetric, hence $MT = M$ .) Pre-multiplying with $M^{-1}$ is not allowed, since $J$ is not of full column rank. However, pre-multiplying with $J^T$ gives:

$$J^T Mt^{ee} = \left(J^T MJ\right)\dot{q} .$$

The matrix $\left(J^T MJ\right)$ is square $(n \times n)$ and full-rank if the manipulator is not in a singular configuration. Hence, it is invertible, and

$$\dot{q} = \left(J^T MJ\right)^{-1} J^T Mt^{ee},$$
$$= J^\dagger_{M} t^{ee} .$$

$J^\dagger_M$ is also a weighted pseudo-inverse of J, but this time with M as the $6 \times 6$ weighting matrix on the space of Cartesian twists. (And not on the joint space of the robot!) Pre-multiplying Eq. $\dot{q} = \left(J^T MJ\right)^{-1} J^T Mt^{ee}$, with $J$ proves that the executed twist t is a projection of the desired twist

$$= J^\dagger_{M} t^{ee} .$$

$t^{ee}$ :
$$t = J\dot{q} = J^\dagger_M t^{ee} = Pt^{ee} , \text{ with } P = JJ^\dagger_M.$$

It is straightforward to check that $P$ indeed satisfies the projection operator property that $P P = P$ . A set of linear constraints similar to Eq. $A(q)\dot{q} = 0$. does not exist: all joint velocities are possible.

## Motion in Contact

Assume the manipulator has six joints, but its end-effector makes contact with a (stiff) environment. This means that it looses a number of degrees of motion freedom, say n. The Jacobian J is still a $6 \times 6$ matrix, but an n-dimensional wrench space exists (with wrench basis G) to which the allowed motions of the end-effector must be reciprocal. G represents the Cartesian directions in which the contact can generate contact forces, so it imposes a set of linear constraints on the joint velocities as in Eq. $A(q)\dot{q} = 0$:

$$\left(G^T \ddot{A} J\right)\dot{q} = 0.$$

This suggests a procedure to "filter" a given end-effector twist $t^{ee} \in \text{span}\left(J\right)$ into a twist t compatible with the constraint (i.e., reciprocal to $G$). Indeed, this "kinetostatic filtering" can be formulated as the following constrained optimization problem:

$$\begin{cases} \min_t T = \frac{1}{2}\left(t^{ee}-t\right)^T M\left(t^{ee}-t\right) \\ \text{such that } G^T \ddot{\mathbf{A}}t = 0 \end{cases}$$

The solution of this optimization problem runs along similar lines (i) include the constraint in the objective function by means of Lagrange multipliers; (ii) set the partial derivative with respect to t equal to zero; and (iii) solve for the Lagrange multipliers. This leads to the following weighted projection operation:

$$t = \left(1-\left(\left(\ddot{\mathbf{A}}G\right)^T\right)^{\dagger}_{M^{-T}}\left(\ddot{\mathbf{A}}G\right)^T\right)t^{ee}.$$

## Cooperating Serial Robots

Two serial robots = equivalent redundant serial robot. Three or more serial robots = tree-structured robot. Cooperation on end-effector level, vs cooperation at link level.

## References

- Robot, definition: searchenterpriseai.techtarget.com, Retrieved 6 June, 2019

- Kurfess, Thomas R. (1 January 2005). Robotics and Automation Handbook. Taylor & Francis. ISBN 9780849318047. Archived from the original on 4 December 2016. Retrieved 5 July 2016 – via Google Books

- What-is-a-robot-4148364: lifewire.com, Retrieved 7 July, 2019

- Beasley, Ryan A. (12 August 2012). "Medical Robots: Current Systems and Research Directions". Journal of Robotics. 2012: 1–14. Doi:10.1155/2012/401613

- The-future, robot-technology, technology: britannica.com, Retrieved 8 Augus,t 2019

- Geoffrey A. Landis, "Exploring Venus by Solar Airplane," Space Technology Applications International Forum; 11–15 Feb. 2001; Albuquerque, NM, AIP Conference Proceedings Vol. 552, pp. 16–18 (NASA NTRS Retrieved 13 May 2015)

- Types-of-robots-based-on-configuration: plantautomation-technology.com, Retrieved 9 January, 2019

- Bekey, George A. (2005). Autonomous robots: from biological inspiration to implementation and control. Cambridge, Massachusetts: MIT Press. ISBN 978-0-262-02578-2

- Israel, Brett (2016-12-06). "Wall-jumping robot is most vertically agile ever built". Berkeley News. Retrieved 2017-06-07

# 3

# Robotics: Components and Processes

There are a number of components and processes used in robotics such as robotic sensing, actuation, robotics simulator, robotic arm, robotic sensors and robotic manipulation. The topics elaborated in this chapter will help in gaining a better perspective about these components and processes of robotics.

The structure of a robot is usually mostly mechanical and is called a kinematic chain (its functionally similar to the skeleton of human body). The chain is formed of links (its bones), actuators (its muscles) and joints which can allow one or more degrees of freedom.

Some robots use open serial chains in which each link connects the one before to the one after it. Robots used as manipulators have an end effectors mounted on the last link. This end effectors can be anything from a welding device to a mechanical hand used to manipulate the environment.

1. Actuation:

Actuation is the "muscles" of a robot, the parts which convert stored energy into movement. The most popular actuators are electric motors.

2. Motors:

The vast majority of robots use electric motors, including bushed and brushies DC motors.

3. Stepper motors:

Stepper motors do not spin freely like DC motors; they rotate in discrete steps, under the command of a controller. This makes them easier to control.

4. Piezo Motors:

A recent alternative to DC Motors are piezo motors or ultrasonic motors. Tiny piezoceramic elements, vibrating many thousands of times per second, cause linear or rotary motion.

5. Air Muscles:

The air muscle in a simple yet powerful device for providing a pulling force. It behaves in a very similar way to a biological muscle; it can be used to construct robots with a similar muscle/skeleton system to an animal.

6. Electroactive polymers:

Are classes of plastics which change shape in response to electric stimulation.

7. Elastic Nanotubes:

The absence of defects in nanotubes enables these filaments to deform elastically by several percent.

8. Manipulation:

Robots work in the real world require some way to manipulate objects; pick up, modify, destroy or otherwise have an effect. Thus, the hands, of a robot are often referred to as end effectors. Most robots arms are replaceable effectors, each allowing them to perform some small range of tasks. Some have a fixed manipulator which cannot be replaced, while a few have one very general purpose manipulator.

# ROBOTIC SENSING

Robotic sensing is a subarea of robotics science intended to give robots sensing capabilities, so that robots are more human-like. Robotic sensing mainly gives robots the ability to see, touch, hear and move and uses algorithms that require environmental feedback.

## Touch

### Signal Processing

Touch sensory signals can be generated by the robot's own movements. It is important to identify only the external tactile signals for accurate operations. Previous solutions employed the Wiener filter, which relies on the prior knowledge of signal statistics that are assumed to be stationary. Recent solution applies an adaptive filter to the robot's logic. It enables the robot to predict the resulting sensor signals of its internal motions, screening these false signals out. The new method improves contact detection and reduces false interpretation.

### Usage

Touch patterns enable robots to interpret human emotions in interactive applications. Four measurable features—force, contact time, repetition, and contact area change—can effectively categorize touch patterns through the temporal decision tree classifier to account for the time delay and associate them to human emotions with up to 83% accuracy. The Consistency Index is applied at the end to evaluate the level of confidence of the system to prevent inconsistent reactions.

Robots use touch signals to map the profile of a surface in hostile environment such as a water pipe. Traditionally, a predetermined path was programmed into the robot. Currently, with the integration of touch sensors, the robots first acquire a random data point; the algorithm of the robot will then determine the ideal position of the next measurement according to a set of predefined geometric primitives. This improves the efficiency by 42%.

In recent years, using touch as a stimulus for interaction has been the subject of much study. In 2010, the robot seal PARO was built, which reacts to many stimuli from human interaction, including touch. The therapeutic benefits of such human-robot interaction is still being studied, but has shown very positive results.

## Hearing

### Signal Processing

Accurate audio sensors require low internal noise contribution. Traditionally, audio sensors combine acoustical arrays and microphones to reduce internal noise level. Recent solutions combine also piezoelectric devices. These passive devices use the piezoelectric effect to transform force to voltage, so that the vibration that is causing the internal noise could be eliminated. On average, internal noise up to about 7dB can be reduced.

Robots may interpret strayed noise as speech instructions. Current voice activity detection (VAD) system uses the complex spectrum circle centroid (CSCC) method and a maximum signal-to-noise ratio (SNR) beamformer. Because humans usually look at their partners when conducting conversations, the VAD system with two microphones enable the robot to locate the instructional speech by comparing the signal strengths of the two microphones. Current system is able to cope with background noise generated by televisions and sounding devices that come from the sides.

### Usage

Robots can perceive emotions through the way we talk. Acoustic and linguistic features are generally used to characterize emotions. The combination of seven acoustic features and four linguistic features improves the recognition performance when compared to using only one set of features.

### Acoustic Feature

- Duration,
- Energy,
- Pitch,
- Spectrum,
- Cepstral,
- Voice quality,
- Wavelets.

### Linguistic Feature

- Bag of words,
- Part-of-speech,

- Higher semantics,

- Varia.

## Movement

### Usage

Automated robots require a guidance system to determine the ideal path to perform its task. However, in the molecular scale, nano-robotslack such guidance system because individual molecules cannot store complex motions and programs. Therefore, the only way to achieve motion in such environment is to replace sensors with chemical reactions. Currently, a molecular spider that has one streptavidin molecule as an inert body and three catalytic legs is able to start, follow, turn and stop when came across different DNA origami. The DNA-based nano-robots can move over 100 nm with a speed of 3 nm/min.

In a TSI operation, which is an effective way to identify tumors and potentially cancer by measuring the distributed pressure at the sensor's contacting surface, excessive force may inflict a damage and have the chance of destroying the tissue. The application of robotic control to determine the ideal path of operation can reduce the maximum forces by 35% and gain a 50% increase in accuracy compared to human doctors.

### Performance

Efficient robotic exploration saves time and resources. The efficiency is measured by optimality and competitiveness. Optimal boundary exploration is possible only when a robot has square sensing area, starts at the boundary, and uses the Manhattan metric. In complicated geometries and settings, a square sensing area is more efficient and can achieve better competitiveness regardless of the metric and of the starting point.

# SOURCE OF POWER

The *InSight* lander with solar panels deployed in a cleanroom.

At present, mostly (lead–acid) batteries are used as a power source. Many different types of batteries can be used as a power source for robots. They range from lead–acid batteries, which are safe and have relatively long shelf lives but are rather heavy compared to silver–cadmium batteries that are much smaller in volume and are currently much more expensive. Designing a battery-powered robot needs to take into account factors such as safety, cycle lifetime and weight. Generators, often some type of internal combustion engine, can also be used. However, such designs are often mechanically complex and need a fuel, require heat dissipation and are relatively heavy. A tether connecting the robot to a power supply would remove the power supply from the robot entirely. This has the advantage of saving weight and space by moving all power generation and storage components elsewhere. However, this design does come with the drawback of constantly having a cable connected to the robot, which can be difficult to manage. Potential power sources could be:

- Pneumatic (compressed gases).

- Solar power (using the sun's energy and converting it into electrical power).

- Hydraulics (liquids).

- Flywheel energy storage.

- Organic garbage (through anaerobic digestion).

- Nuclear.

Power sources are indispensable while designing robotic systems. Hence, the selection of power sources should be the primary focus owing to its impact on the mechanism, packaging, weight and size of the system.

At present, batteries are more commonly used power sources. Many different types of batteries ranging from lead acid batteries that are safe to silver cadmium batteries that are smaller in volume.

Weight of the robot, cycle lifetime and safety are the factors that need to be taken into account while designing a battery-powered robot.

The following sections will elaborate on some of the important power components employed in robotic systems.

## Generator Systems

The generator systems work based on the conversion of the gasoline energy to motive power via a combustion process in an engine.

These are two stage systems that can charge the robotic devices continuously during operation, even when the robots are immobile.

To the contrary, thermoelectric generators directly convert heat into electricity through the Seebeck effect. This type of power generation can be scaled to very small sizes with affecting the energy efficiency.

## Hybrid Systems

Hybrid/two-stage power systems are the system in which the power is continuously supplied at the first stage, and power is delivered as and when needed at the second stage.

Hybrid systems are of two types: serial and parallel. In series hybrid systems, a small amount of energy is continuously produced at the first stage, and the second stage supplies high power for short durations. In parallel configuration, power can be drawn at any time.

## Batteries

Batteries are the main component of a robotic system. Batteries can be classified into rechargeable or non-rechargeable. Non-rechargeable batteries deliver more power based on their size, and are suitable for certain applications.

Alkaline batteries are inexpensive, and lithium batteries, on the other hand exhibit a longer shelf life and better performance.

Common rechargeable batteries such as nickel-cadmium (NiCd) and lead acid batteries deliver a smaller voltage than alkaline batteries. They are found in battery packs along with specialized power connectors. Gelled lead acid batteries are widely used and capable of providing power of up to 40Wh/kg.

Lithium-ion, nickel metal hydride and silver zinc batteries are some of the other rechargeable battery technologies that offer significantly increased energy density.

## Photovoltaic Cells

Photovoltaic or solar cells can be used to charge the batteries of the robotic systems. They are used in conjunction with a capacitor and can be charged up to a set voltage level and then be discharged via the movements of motor. These cells are chiefly used in BEAM robots.

## Fuel Cells

Like batteries, fuel cells supply direct energy via a non-combustion process by directly deriving power from a hydrocarbon source at high efficiencies of up to 75%. This includes two electrodes sandwiched around a conductive electrolyte.

The electrons are released from the anode in the presence of a platinum catalyst, and they are used to generate an electrical current through a load. The efficiency of fuel cells can be increased to nearly 80% by utilizing the waste heat.

Other potential power sources of robotic systems include:

- Flywheel energy storage.
- Hydraulics.
- Compressed gases.
- Super capacitors.
- Organic garbage.

# ACTUATION

Actuators, also known as drives, are mechanisms for getting robots to move. Most actuators are powered by pneumatics (air pressure), hydraulics (fluid pressure), or motors (electric current). Most actuation uses electromagnetic motors and gears but there have been frequent uses of other forms of actuation including NiTinOL "muscle-wires" and inexpensive Radio Control servos. To get a motor under computer control, different motor types and actuator types are used. Some of the motor types are Synchronous, Stepper, AC servo, Brushless DC servo, and Brushed DC servo. Radio Control servos for model airplanes, cars and other vehicles are light, rugged, cheap and fairly easy to interface. Some of the units can provide very high torque speed. A Radio Control servo can be controlled from a parallel port. With one of the PCs internal timers cranked up, it is possible to control eight servos from a common parallel port with nothing but a simple interrupt service routine and a cable. In fact, power can be pulled from the disk drive power connector and the PC can run all servos directly with no additional hardware. The only down side is that the PC wastes some processing power servicing the interrupt handler.

## DC Motors

The most common actuator you will use (and the most common in mobile robotics in general) is the direct current (DC) motor. They are simple, cheap, and easy to use. Also, they come in a great variety of sizes, to accommodate different robots and tasks. DC motors convert electrical into mechanical energy. They consist of permanent magnets and loops of wire inside. When current is applied, the wire loops generate a magnetic field, which reacts against the outside field of the static magnets. The interaction of the fields produces the movement of the shaft/armature. Thus, electromagnetic energy becomes motion. As with any physical system, DC motors are not perfectly efficient, meaning that the energy is not converted perfectly, without any waste. Some energy is wasted as heat generated by friction of mechanical parts. Inefficiencies are minimized in well-designed (and more expensive) motors, and their performance can be brought up to the 90th percentile, but cheap motors (such as the ones you may use) can be as low as 50%. (In case you think this is very inefficient, remember that other types of effectors, such as miniature electrostatic motors, may have much lower efficiencies still.) A motor requires a power source within its operating voltage, i.e., the recommended voltage range for best efficiency of the motor. Lower voltages will usually turn the motor (but provide less power). Higher voltages are more tricky: in some cases they can increase the power output but almost always at the expense of the operating life of the motor. E.g., the more you rev your car engine, the sooner it will die. When constant voltage is applied, a DC motor draws current in the amount proportional to the work it is doing. For example, if a robot is pushing against a wall, it is drawing more current (and draining more of its batteries) than when it is moving freely in open space.

The reason is the resistance to the motor motion introduced by the wall. If the resistance is very high (i.e., the wall just won't move no matter how much the robot pushes against it), the motor draws a maximum amount of power, and stalls. This is defined as the stall current of the motor: the most current it can draw at its specified voltage. Within a motor's operating current range, the more current is used, the more torque or rotational force is produced at the shaft. In general, the strengths of the magnetic field generated in the wire loops is directly proportional to the applied current and thus the produced torque at the shaft. Besides stall current, a motor also has its stall torque, the amount of rotational force produced when the motor is stalled at its operating voltage. Finally, the amount of power a motor generates is the product of its shaft's rotational velocity and its torque. If there is no load on the shaft, i.e., the motor is spinning freely, then the rotational velocity is the highest, but the torque is 0, since no mechanism is being driven by the motor. The output power, then, is 0 also. In contrast, when the motor is stalled, it is producing maximum torque, but the rotational velocity is 0, so the output power is 0 again.

Between free spinning and stalling, the motor does useful work, and the produced power has a characteristic parabolic relationship demonstrating that the motor produces the most power in the middle of its performance range. Most DC motors have unloaded speeds in the range of 3,000 to 9,000 RPM (revolutions per minute), or 50 to 150 RPS (revolutions per second). That turns out to put them in the high-speed but low-torque category (compared to some other alternatives). For example, how often do you need to drive something very light that rotates very fast (besides a fan)? Yet that is what DC motors are naturally best at. In contrast, robots need to pull loads (i.e., move their bodies and manipulators, all of which have significant mass), thus requiring more torque and less speed. As a result, the performance of a DC motor typically needs to be adjusted in that direction, through the use of gears.

## Gearing

The force generated at the edge of a gear is equal to the product of the radius of the gear and its torque (F = r t), in the line tangential to its circumference. By combining gears with different radii, we can manipulate the amount of force/torque the mechanism generates. The relationship between the radii and the resulting torque is well defined, as follows: Suppose Gear1 with radius $r_1$ turns with torque $t_1$, generating a force of $t_1/r_1$ perpendicular to its circumference. Now if we mesh it with Gear2, with r2, which generates t2/r2, then t1/r1 = t2/r2. To get the torque generated by Gear2, we get: t2 = t1 r2/r1. Intuitively, this means: the torque generated at the output gear is proportional to the torque on the input gear and the ratio of the two gear's radii. If r2 > r1, we get a bigger number, if r1 > r2, we get a smaller number.

If the output gear is larger than the input gear, the torque increases. If the output gear is smaller than the input gear, the torque decreases. Besides the change in torque that takes place when gears are combined, there is also a corresponding change in speed. To measure speed we are interested in the circumference of the gear, C= 2 * pi * r. Simply put, if the circumference of Gear1 is twice that of Gear2, then Gear2 must turn twice for each full rotation of Gear1. If the output gear is larger than the input gear, the speed decreases. If the output gear is smaller than the input gear, the speed increases. In summary, when a small gear drives a large one, torque is increased and speed is decreased. Analogously, when a large gear drives a small one, torque is decreased and speed is increased. Thus, gears are used in DC motors (which we said are fast and low torque) to trade off extra speed for additional torque. Gears are combined using their teeth. The number of teeth is not arbitrary, since it is the key means of proper reduction. Gear teeth require special design so that they mesh properly. If there is any looseness between meshing gears, this is called backlash, the ability for a mechanism to move back \& forth within the teeth, without turning the whole gear.

Reducing backlash requires tight meshing between the gear teeth, but that, in turn, increases friction. As you can imagine, proper gear design and manufacturing is complicated. To achieve "three to one gear reduction (3:1)", we apply power to a small gear (say one with 8-teeth) meshed with a large one (with 3 * 8 = 24 teeth). As a result, we have slowed down the large gear by 3 and have tripled its torque. Gears can be organized in series ("ganged"), in order to multiply their effect. For example, 2 3:1 gears in series result in a 9:1 reduction. This requires a clever arrangement of gears. Or three 3:1 gears in series can produce a 27:1 reduction. This method of multiplying reduction is the underlying mechanism that makes DC motors useful and ubiquitous.

## Electronic Control of Motors

It should come as no surprise that motors require more battery power (i.e., more current) than electronics (e.g., 5 milliamps for the 68HC11 processor v. 100 milliamps - 1 amp for a small DC motor). Typically, specialized circuitry is required. You need to learn about H-bridges and pulse-width modulation there.

## Servo Motors

It is sometimes necessary to be able to move a motor to a specific position. If you consider your basic DC motor, it is not built for this purpose. Motors that can turn to a specific position are called servo motors and are in fact constructed out of basic DC motors, by adding:

- Some gear reduction.

- A position sensor for the motor shaft.

- An electronic circuit that controls the motor's operation.

Servos are used in toys a great deal, to adjust steering on steering in RC cars and wing position in RC airplanes. Since positioning of the shaft is what servo motors are all about, most have their movement reduced to 180 degrees. The motor is driven with a waveform that specifies the desired angular position of the shaft within that range. The waveform is given as a series of pulses, within a pulse-width modulatedsignal. Thus, the width (i.e., length) of the pulse specifies the control value for the motor, i.e., how the shaft should turn. Therefore, the exact width/length of the pulse is critical, and cannot be sloppy. There are no milliseconds or even microseconds to be wasted here, or the motor will behave very badly, jitter, and go beyond its mechanical limit. This limit should be checked empirically, and avoided. In contrast, the duration between the pulses is not critical at all. It should be consistent, but there can be noise on the order of milliseconds without any problems for the motor. This is intuitive: when no pulse arrives, the motor does not move, so it simply stops. As long as the pulse gives the motor sufficient time to turn to the proper position, additional time does not hurt it.

Solar speeder.

## Continuous Rotation Motors

A regular DC motor can be used for continuous rotation. Furthermore, servo motors can also be retrofitted to provide continuous rotation (remember, they only to 180 otherwise), like this:

- Remove mechanical limit (revert back to DC motor shaft).

- Remove pot position sensor (no need to tell position).

- Apply 2 resistors to fool the servo to think it is fully turning.

Actuators are the "muscles" of a robot, the parts which convert stored energy into movement. By far the most popular actuators are electric motors that rotate a wheel or gear, and linear actuators that control industrial robots in factories. There are some recent advances in alternative types of actuators, powered by electricity, chemicals, or compressed air.

## Electric Motors

The vast majority of robots use electric motors, often brushed and brushless DC motors in portable robots or AC motors in industrial robots and CNC machines. These motors are often preferred in systems with lighter loads, and where the predominant form of motion is rotational.

## Linear Actuators

Various types of linear actuators move in and out instead of by spinning, and often have quicker direction changes, particularly when very large forces are needed such as with industrial robotics. They are typically powered by compressed and oxidized air (pneumatic actuator) or an oil (hydraulic actuator) Linear actuators can also be powered by electricity which usually consists of a motor and a leadscrew. Another common type is a mechanical linear actuator that is turned by hand, such as a rack and pinion on a car.

## Series Elastic Actuators

A flexure is designed as part of the motor actuator, to improve safety and provide robust force control, energy efficiency, shock absorption (mechanical filtering) while reducing excessive wear on the transmission and other mechanical components. The resultant lower reflected inertia can improve safety when a robot is interacting with humans or during collisions. It has been used in various robots, particularly advanced manufacturing robots and walking humanoid robots.

## Air Muscles

Pneumatic artificial muscles, also known as air muscles, are special tubes that expand(typically up to 40%) when air is forced inside them. They are used in some robot applications.

## Muscle Wire

Muscle wire, also known as shape memory alloy, Nitinol® or Flexinol® wire, is a material which contracts (under 5%) when electricity is applied. They have been used for some small robot applications.

## Electroactive Polymers

EAPs or EPAMs are a plastic material that can contract substantially (up to 380% activation strain) from electricity, and have been used in facial muscles and arms of humanoid robots, and to enable new robots to float, fly, swim or walk.

## Piezo Motors

Recent alternatives to DC motors are piezo motors or ultrasonic motors. These work on a fundamentally different principle, whereby tiny piezoceramic elements, vibrating many thousands of times per second, cause linear or rotary motion. There are different mechanisms of operation; one type uses the vibration of the piezo elements to step the motor in a circle or a straight line. Another type uses the piezo elements to cause a nut to vibrate or to drive a screw. The advantages of these motors are nanometer resolution, speed, and available force for their size. These motors are already available commercially, and being used on some robots.

## Elastic Nanotubes

Elastic nanotubes are a promising artificial muscle technology in early-stage experimental development. The absence of defects in carbon nanotubes enables these filaments to deform elastically by several percent, with energy storage levels of perhaps 10 J/cm3 for metal nanotubes. Human biceps could be replaced with an 8 mm diameter wire of this material. Such compact "muscle" might allow future robots to outrun and outjump humans.

# ROBOTICS SIMULATOR

Robologix robotics simulator.

A robotics simulator is used to create application for a physical robot without depending on the actual machine, thus saving cost and time. In some case, these applications can be transferred onto the physical robot (or rebuilt) without modifications.

The term *robotics simulator* can refer to several different robotics simulation applications. For example, in mobile robotics applications, behavior-based robotics simulators allow users to create simple worlds of rigid objects and light sources and to program robots to interact with these worlds. Behavior-based simulation allows for actions that are more biological in nature when compared to simulators that are more binary, or computational. In addition, behavior-based simulators may "learn" from mistakes and are capable of demonstrating the anthropomorphic quality of tenacity.

One of the most popular applications for robotics simulators is for 3D modeling and rendering of a robot and its environment. This type of robotics software has a simulator that is a virtual robot, which is capable of emulating the motion of an actual robot in a real work envelope. Some robotics simulators use a physics engine for more realistic motion generation of the robot. The use of a robotics simulator for development of a robotics control program is highly recommended regardless of whether an actual robot is available or not. The simulator allows for robotics programs to be conveniently written and debugged off-line with the final version of the program tested on an actual robot. This primarily holds for industrial robotic applications only, since the success of off-line programming depends on how similar the real environment of the robot is to the simulated environment.

*Sensor-based* robot actions are much more difficult to simulate and/or to program off-line, since the robot motion depends on the instantaneous sensor readings in the real world.

## Features

Modern simulators tend to provide the following features:

- Fast robot prototyping:

  - Using the own simulator as creation tool (Virtual Robot Experimentation Platform, Webots, R-Station, Marilou, 4DV-Sim).

    ∘   Using external tools.

- Physics engines for realistic movements. Most simulators use ODE (Gazebo, LpzRobots, Marilou, Webots) or PhysX (Microsoft Robotics Studio, 4DV-Sim).

- Realistic 3d rendering. Standard 3d modeling tools or third party tools can be used to build the environments.

- Dynamic robot bodies with scripting. C, C++, Perl, Python, Java, URBI, MATLAB languages used by Webots, Python used by Gazebo.

## Simulators

Among the newest technologies available today for programming are those which use a virtual simulation. Simulations with the use of virtual models of the working environment and the robots themselves can offer advantages to both the company and programmer. By using a simulation, costs are reduced, and robots can be programmed off-line which eliminates any down-time for an assembly line. Robot actions and assembly parts can be visualised in a 3-dimensional virtual environment months before prototypes are even produced. Writing code for a simulation is also easier than writing code for a physical robot. While the move toward virtual simulations for programming robots is a step forward in user interface design, many such applications are only in their infancy.

## Technical Information

| Software | Main programming language | Formats support | Extensibility | External APIs | Robotics middleware support | Primary user interface | Headless simulation |
|---|---|---|---|---|---|---|---|
| Actin | C++ | SLDPRT, SLDASM, STEP, OBJ, STL, 3DS, Collada, VRML, URDF, XML, ECD, ECP, ECW, ECX, ECZ, | Plugins (C++), API | Unknown | ROS | GUI | Yes (Act-inRT) |
| ARS | Python | Unknown | Python | Unknown | None | Unknown | Unknown |
| AUTO-MAPPPS | C++, Python | STEP, IGES, STL | Plugins (C), API | XML, C | Socket | GUI | Yes |
| Gazebo | C++ | SDF/URDF, OBJ, STL, Collada | Plugins (C++) | C++ | ROS, Player, Sockets (protobuf messages) | GUI | Yes |
| MORSE | Python | Unknown | Python | Python | Sockets, YARP, ROS, Pocolibs, MOOS | Command-line | Yes |
| OpenHRP | C++ | VRML | Plugins (C++), API | C/C++, Python, Java | Open-RTM-aist | GUI | Unknown |
| RoboDK | Python | SLDPRT, SLDASM, STEP, OBJ, STL, 3DS, Collada, VRML, URDF, Rhinoceros_3D, ... | API, Plug-In Interface | Python, C/C++, C#, Matlab, ... | Socket | GUI | Unknown |

| Software | Main programming language | Formats support | Extensibility | External APIs | Robotics middleware support | Primary user interface | Headless simulation |
|---|---|---|---|---|---|---|---|
| SimSpark | C++, Ruby | Ruby Scene Graphs | Mods (C++) | Network (sexpr) | Sockets (sexpr) | GUI, Sockets | Unknown |
| V-Rep | LUA | OBJ, STL, DXF, 3DS, Collada, URDF | API, Add-ons, Plugins | C/C++, Python, Java, Urbi, Matlab/Octave | Sockets, ROS | GUI | Yes |
| Webots | C++ | WBT, VRML, X3D | API, PROTOs, Plugins (C/C++) | C, C++, Python, Java, Matlab, ROS | Sockets, ROS, NaoQI | GUI | Yes |
| 4DV-Sim | C++ | 3DS, OBJ, Mesh | Plugins (C++), API | FMI/FMU, Matlab | ROS, Sockets, Plug & Play interfaces | GUI | Yes |
| OpenRAVE | C++, Python | XML, VRML, OBJ, Collada | Plugins (C++), API | C/C++, Python, Matlab | Sockets, ROS, YARP | GUI, Sockets | Yes |
| Software | Main programming language | Formats support | Extensibility | External APIs | Robotic middleware support | Primary user interface | Headless simulation |

## Robotic Simulators

## Open Source Simulators

- Breve: a 3D-world multi-agent simulator in Python.

- EZPhysics: Combination of Ogre3D and ODE physics, GUI exposes all of ODE's objects data, network closed loop remote control optionally via Matlab/Simulink.

- Gazebo Simulator: An open-source 3D robotics simulator used in a number of DARPA contests.

- Khepera Simulator an open-source Windows simulator for the Khepera robot predating Webots.

- Klamp't: a simulator introduced in 2013 specializing in stable trimesh-trimesh contact. Supports legged locomotion and manipulation.

- LpzRobots: a 3D-physics robot simulator developed at the University of Leipzig.

- MiniBloq: This robot programing software for Arduino boards has a new simulator.

- Moby: a rigid-body dynamics library written in C++.

- OpenSim Simulator for articulated and wheeled robots with a wide range of characteristics. Further development stopped in the year 2006.

- Robotics Toolbox for MATLAB is Free Software that provides functionality for representing pose (homogeneous transformations, Euler and RPY angles, quaternions), arm robots (forward/inverse kinematics, dynamics, simulation, animation) and mobile robots (control, localisation, planning and animation).

- ARTE A Robotics Toolbox for Education (ARTE) is a Free Software educational tool based on Matlab. It provides functions to represent position and orientation. As well, includes functions to simulate robotic arms (direct/inverse kinematics, dynamics, path planning and more). The toolbox includes a large set of 3D robotic models that can be viewed and simulated inside a robotic cell.

- Simbad 3d Robot Simulator Java based simulator.

- SimRobot: A robot simulator software package developed at the Universität Bremen and the German Research Center for Artificial Intelligence.

- Stage: 2.5D simulator often used with Player to form the Player/Stage system. Part of the Player Project.

- STDR Simulator A simple, flexible and scalable 2D multi-robot simulator for use within Robot Operating System.

- UCHILSIM: A physics based simulator for AIBO Robots introduced in RoboCup 2004.

- UWSim : an UnderWater SIMulator for marine robotics research and development which incorporates sensor, dynamic and physics simulation.

## Closed-Source and Proprietary Simulators

- AnyKode Marilou

- ORCA-Sim: a (Windows) 3D robot simulator using the Newton Game Dynamics physics engine.

## Autonomous Vehicle Simulator

## Open Source Simulators

- Carla is an open-source simulator for autonomous driving research. CARLA provides open digital assets (urban layouts, buildings, vehicles) that were created for this purpose and can be used freely. The simulation platform supports flexible specification of sensor suites and environmental conditions.

- AirSim AirSim is a simulator for drones, cars and more, built on Unreal Engine. It is open-source, cross platform, and supports hardware-in-loop with popular flight controllers such as PX4 for physically and visually realistic simulations.

## Closed-source and Proprietary Simulators

- LGSVL is a unity-based multi-robot simulator for autonomous vehicle developers.

# ROBOTIC ARM

Most common manufacturing robot is the Robotic Arm, which is a mechanical arm, usually programmable which perform the same function or similar to the human arm. The robotic arm usually made up of seven metal segment, joined by six joints. An industrial robot with six joints closely resembles a human arm — it has the equivalent of a shoulder, an elbow, and a wrist.

## Types of Robotic Arm

The robotic arm can be characterized into 5 major categories based on its mechanical structure.

- Cartesian Robot: it has three joints which coincide with standard X-Y-Z Cartesian axes.

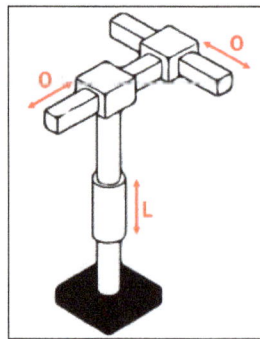

Image of a Cartesian robot arm.

- Cylindrical robot: it has a number of joints that rotate on cylindrical axes, which rotate on one fixed rod.

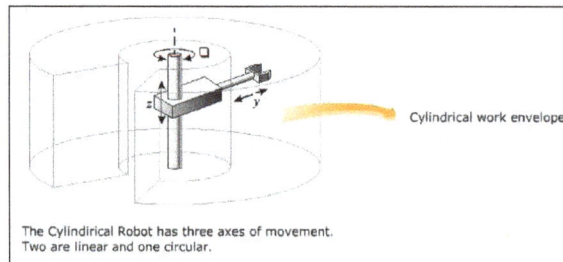

Image of a cylindrical robot arm.

- Spherical or polar arm: they are those with joints that allow it full rotation throughout a spherical Range.

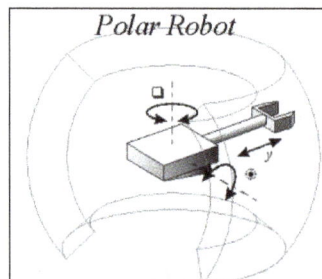

Image of a polar robot arm.

- SCARA Robot: These robots have two parallel rotary joints to allow full movement throughout a plane, typically for pick-and-place work.

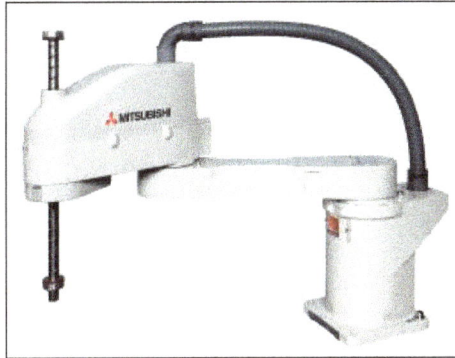

Image of a SCARA Robot.

- Articulated Robots: most commonly used configuration because of their flexibility in reaching any part of the working envelope. Mostly used in such complex application as welding, drilling and soldering operations.

Image of an Articulated Robot.

## Application

Robotic arms are typically used for industrial applications. The application is the type of work robot is designed to do.

Different applications will have different requirements. For example, a painting robot will require a small payload but a large movement range. On the other hand, an assembly robot will have a small workspace but will be very precise and fast. Industrial robots are designed for specific applications and based on their function will have their own movement, linkage dimension, control law, software, and accessory packages. Below are some types of applications:

- Welding Robot.
- Material Handling Robot.

- Palletizing Robot.

- Painting robot.

- Assembly Robot.

## Future Outlook

There are some special types of robotic arms that can be brought into our workspace in the near future. Some of them are as follows:

- Dobot Robotic arm: it is a 4-axis robotic arm designed for makers, artist, educators, and scientist. The arm can be controlled with the help of an APP by phone or PC via Bluetooth.

Dobot Robotic Arm.

- Make arm: It is mostly used in 3D printers, laser cut and engrave, carve and mill, writes, and plots, PCB fabrication, assembly, picks & place, etc. It can be used as our "personal fabrication system "that can mount on your desktop.

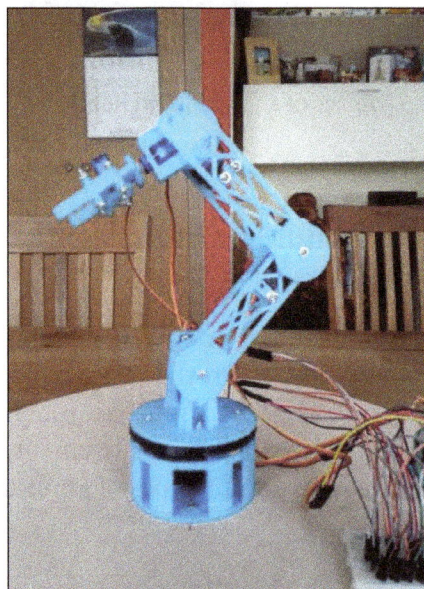
Make arm robot.

- Collaborative robot: they are manufactured in such a way that they follow some safety standard so that they cannot hurt the human.

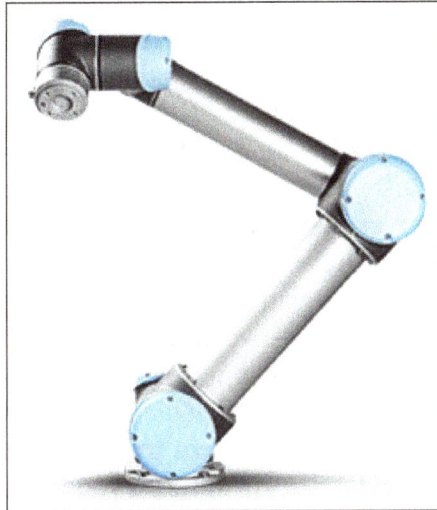
Sample Collaborative robot arm.

## Robot End Effector

In robotics, an end effector is the device at the end of a robotic arm, designed to interact with the environment. The exact nature of this device depends on the application of the robot.

In the strict definition, which originates from serial robotic manipulators, the end effector means the last link (or end) of the robot. At this endpoint the tools are attached. In a wider sense, an end effector can be seen as the part of a robot that interacts with the work environment. This does not refer to the wheels of a mobile robot or the feet of a humanoid robot which are also not end effectors—they are part of the robot's mobility.

End effectors may consist of a gripper or a tool. When referring to robotic prehension there are four general categories of robot grippers, these are:

- Impactive – jaws or claws which physically grasp by direct impact upon the object.

- Ingressive – pins, needles or hackles which physically penetrate the surface of the object (used in textile, carbon and glass fibre handling).

- Astrictive – attractive forces applied to the objects surface (whether by vacuum, magneto- or electroadhesion).

- Contigutive – requiring direct contact for adhesion to take place (such as glue, surface tension or freezing).

They are based on different physical effects used to guarantee a stable grasping between a gripper and the object to be grasped. Industrial grippers can be mechanical, the most diffused in industry, but also based on suction or on the magnetic force. Vacuum cups and electromagnets dominate the automotive field and in particular metal sheet handling. Bernoulli grippers exploit the airflow between the gripper and the part that causes a lifting force which brings the gripper and part close each other (i.e. the Bernoulli's principle). Bernoulli grippers are a type of contactless grippers, namely the object remains confined in the force field generated by the gripper without coming into direct contact with it. Bernoulli grippers have been adopted in photovoltaic cell handling, silicon waferhandling, and also in the textile and leather industries. Other principles are less used at the macro scale (part size >5mm), but in the last ten years they demonstrated interesting applications in micro-handling. Some of them are ready of spreading out their original field. The other adopted principles are: Electrostatic grippers and van der Waals grippers based on electrostatic charges (i.e. van der Waals' force), capillary grippers and cryogenic grippers, based on liquid medium, and ultrasonic grippers and laser grippers, two contactless grasping principles. Electrostatic grippers are based on charge difference between the gripper and the part (i.e. electrostatic force) often activated by the gripper itself, while van der Waals grippers are based on the low force (still electrostatic) due to the atomic attraction between the molecules of the gripper and those of the object. Capillary grippers use the surface tension of a liquid meniscus between the gripper and the part to center, align and grasp the part, cryogenic grippers freeze a small amount of liquid and the resulting ice guarantees the necessary force to lift and handle the object (this principle is used also in food handling and in textile grasping). Even more complex are ultrasonic based grippers, where pressure standing waves are used to lift up a part and trap it at a certain level (example of levitation are both at the micro level, in screw and gasket handling, and at the macro scale, in solar cell or silicon wafer handling), and laser source that produces a pressure able to trap and move microparts in a liquid medium (mainly cells). The laser gripper are known also as laser tweezers.

A particular category of friction/jaw gripper are the needle grippers: they are called intrusive grippers and exploits both friction and form closure as standard mechanical grippers.

The most known mechanical gripper can be of two, three or even five fingers.

The end effectors that can be used as tools serve various purposes, such as spot welding in an assembly, spray painting where uniformity of painting is necessary, and for other purposes where the working conditions are dangerous for human beings. Surgical robots have end effectors that are specifically manufactured for the purpose.

## Gripper Mechanism

A common form of robotic grasping is force closure.

Generally, the gripping mechanism is done by the grippers or mechanical fingers. Generally only two-finger grippers are used for industrial robots as they tend to be built for specific tasks and can therefore be less complex.

The fingers are also replaceable whether or not the gripper itself is replaced. There are two mechanisms of gripping the object in between the fingers (for the sake of simplicity, the following explanations consider only two finger grippers).

## Shape of the Gripping Surface

The shape of the gripping surface of the fingers can be chosen according to the shape of the objects that are to be manipulated. For example, if a robot is designed to lift a round object, the gripper surface shape can be a concave impression of it to make the grip efficient, or for a square shape the surface can be a plane.

## Force Required to Grip the Object

Though there are numerous forces acting over the body that has been lifted by the robotic arm, the main force acting there is the frictional force. The gripping surface can be made of a soft material with high coefficient of friction so that the surface of the object is not damaged. The robotic gripper must withstand not only the weight of the object but also acceleration and the motion that is caused due to frequent movement of the object. To find out the force required to grip the object, the following formula is used:

$$F = \frac{ma}{\mu n}$$

where:

| $F$ | is | the force required to grip the object, |
|-----|-----|-----|
| $m$ | is | the mass of the object, |
| $a$ | is | the acceleration of the object, |
| $\mu$ | is | the coeffecient of friction, |
| $n$ | is | the number of fingers in the gripper. |

But the above equation is incomplete. The direction of the movement also plays an important role over the gripping of the object. For example, when the body is moved upwards, against the gravitational force, the force required will be more than towards the gravitational force. Hence, another term is introduced and the formula becomes:

$$F = \frac{m(a+g)}{\mu n}$$

Here, the value of $g$ should be taken as the acceleration due to gravity and $a$ the acceleration due to movement.

For many physically interactive manipulation tasks, such as writing and handling a screwdriver, a task-related grasp criterion can be applied in order to choose grasps that are most appropriate to meeting specific task requirements. Several task-oriented grasp quality metrics were proposed to guide the selection of a good grasp that would satisfy the task requirements.

## Examples

The end effector of an assembly line robot would typically be a welding head, or a paint spray gun. A surgical robot's end effector could be a scalpel or others tools used in surgery. Other possible end effectors are machine tools, like a drill or milling cutters. The end effector on the space shuttle's robotic arm uses a pattern of wires which close like the aperture of a camera around a handle or other grasping point.

## Examples of End Effectors

An example of a basic force-closure end effector.

A spot welding end effector.

A repair and observation end effector in use in space (Canadarm2 Latching End Effector).

A highly sophisticated attempt at reproducing the human-hand force-closure end effector.

# ROBOTIC MANIPULATION

Most robots manipulate objects by pick-and-place. There is good reason for this: once a firm grasp is established, the robot can reliably control the motion of the part without needing to continuously sense the state of the part or correct for modeling uncertainties. Most manipulation primitives mentioned above are more sensitive to uncertainties in part state, geometry, mass, friction, and restitution, and to the robot's own control errors. Nonetheless, restricting robots to only grasp objects artificially limits the set of tasks that they can accomplish. Dynamic nonprehensile manipulation raises challenges in high-speed sensing and control, but the dynamics can be exploited to help the robot control object motions that would otherwise be impossible. Leveraging a larger set of manipulation primitives is crucial for robots to reach their full potential in industrial automation, exploration, home care, military, and space applications.

## In-hand Sliding Manipulation

This work focuses on planning manipulation tasks that involve an object sliding in a manipulator's grasp. There are many situations where relative motion between the part and manipulator are useful for a grasp. Sliding can allow for error-corrective motions when performing robot assembly tasks which can decrease the chance of jamming and improve the robustness of planned motions. Sliding can also be used to quickly regrasp objects, and this can be accomplished using external contacts with the environment (e.g. pushing the object against a fixed surface), or with dynamic loads generated by quickly accelerating the manipulator.

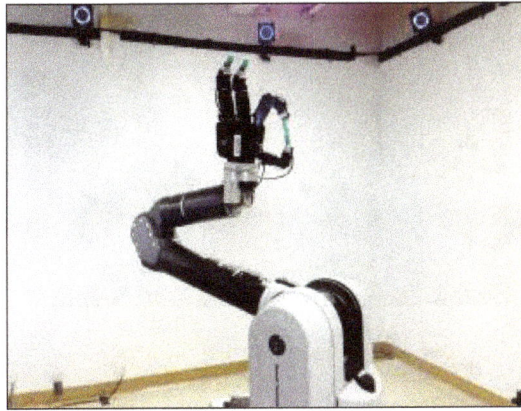

## Hybrid Manipulation Planning and Control

This work focuses on motion planning and control for robotic manipulation tasks in which the manipulator, object, and the environment transition between different contact modes. The dynamic equations that govern how the system evolves over time depend on manipulator controls, contact locations, and whether contacts are fixed/rolling or sliding. This is because the coupling of the manipulator controls to the object through the contacts, and the possible contact forces applied by the environment, are different. We define manipulation primitives according to the number and types of contacts the object makes with a robot and its (rigid) environment. We are currently working on methods to identify different contact primitives, plan between them, and stabilize motion plans using feedback control.

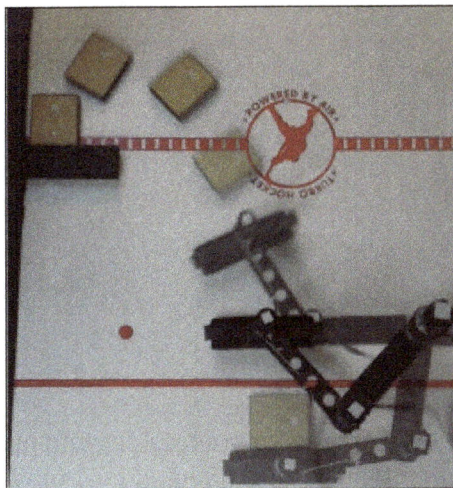

## Vibratory Manipulation

This project examines a very simple and versatile robotic manipulator with surprising capabilities: a six-degree-of-freedom (6-dof) rigid vibrating plate whose motion can be programmed. Even though this manipulator has no grasping ability, it can be used to create programmable vector fields describing how parts will slide on the surface under the influence of friction. These vector fields can be used to manipulate a single part (e.g., to orient or position it), or to manipulate multiple parts along independent trajectories (e.g., to assemble or sort them).

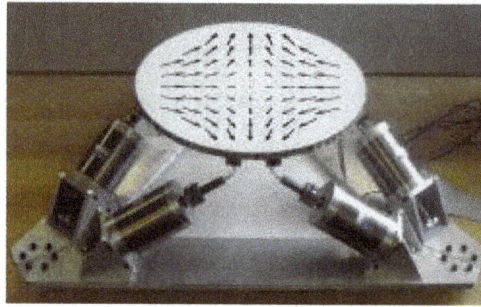

## Rolling Manipulation

Our long-term goal is to develop a unified framework for planning and control of dynamic robotic manipulation. A typical manipulation plan consists of a sequence of manipulation primitives chosen from a library of primitives, with each primitive equipped with its own feedback controller. Problems of interest include planning the motion of the manipulator to achieve the desired motion of the object and feedback control to stabilize the desired trajectory. As a first step to understand the nature of dynamic nonprehensile manipulation, we study feedback stabilization of a canonical rolling problem: balancing a disk-shaped object on top of a disk-shaped manipulator (referred to as the hand) in a vertical plane.

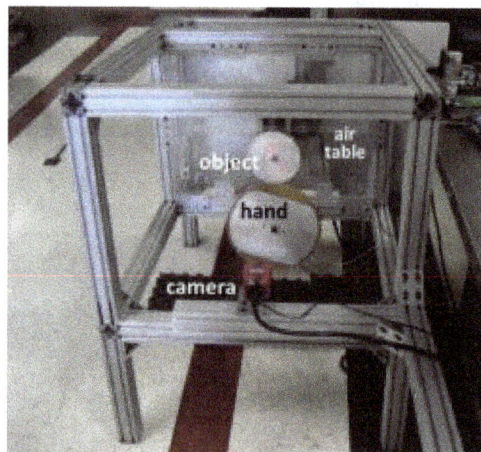

## Manipulator (Device)

Industry-specific robots perform several tasks such as picking and placing objects, movement adapted from observing how similar manual tasks are handled by a fully-functioning human arm. Such robotic arms are also known as robotic manipulators. These manipulators were originally used for applications with respect to bio-hazardous or radioactive materials or for use in inaccessible places.

A series of sliding or jointed segments are put together to form an arm-like manipulator that is capable of automatically moving objects within a given number of degrees of freedom. Every commercial robot manipulator includes a controller and a manipulator arm. The performance of the manipulator depends on its speed, payload weight and precision. However, the reach of its end-effectors, the overall working space and the orientation of the work is determined by the structure of the manipulator.

Robotic manipulators are not a new idea and have already been in operation in many areas of manufacturing for several years. As AI advances, increasing the accuracy and function of robotics will allow a greater range of tasks to be achieved by such manipulators, thus providing more function and provision of labor than the robotic manipulators of the past decades.

## Kinematics of a Robotic Manipulator

A robot manipulator is constructed using rigid links connected by joints with one fixed end and one free end to perform a given task (e.g., to move a box from one location to the next). The joints to this robotic manipulator are the movable components, which enables relative motion between the adjoining links. There are also two linear joints to this robotic manipulator that ensure non-rotational motion between the links, and three rotary type joints that ensure relative rotational motion between the adjacent links.

The manipulator can be divided into two parts, each having different functions:

- Arm and Body – The arm and body of the robot consists of three joints connected together by large links. They can be used to move and place objects or tools within the workspace.

- Wrist – The function of the wrist is to arrange the objects or tools at the work space. Them structural characteristics of the robotic wrist includes two or three compact joints.

## Robotic Manipulator Arm Configuration

Manipulators are grouped into several types based on the combination of joints, which are as follows:

- Cartesian geometry arm – This arm employs prismatic joints to reach any position within its rectangular workspace by using Cartesian motions of the links.

- Cylindrical geometry arm – This arm is formed by the replacement of the waist joint of the Cartesian arm with a revolute joint. It can be extended to any point within its cylindrical workspace by using a combination of translation and rotation.

- Polar/spherical geometry arm – When a shoulder joint of the Cartesian arm is replaced by a revolute joint, a polar geometry arm is formed. The positions of end-effectors of this arm are described using polar coordinates.

- Articulated/revolute geometry arm - Replacing the elbow joint of the Cartesian arm with the revolute joint forms an articulated arm that works in a complex thick-walled spherical shell.

- Selective compliance automatic robot arm (SCARA) – This arm has two revolute joints in a horizontal plane, which allow the arm to extend within a horizontal planar workspace. The TH650A SCARA Robot by TM Robotics is a great example to demonstrate pick and place functionality of robotic manipulators:

## Wrist Configuration

The two main types of wrist design include:

- Roll-pitch-roll or spherical wrist.

- Pitch-yaw-roll.

The spherical wrist is more common because of its mechanically simpler design. It has 6 degrees of freedom, and consists of a Hooke shoulder joint followed by a rotary elbow joint.

## Applications

Some of the major applications of robotic manipulator are discussed below:

- Motion planning,
- Remote handling,
- Micro-robots,
- Humanoid robots,
- Machine tools.

# ROBOTIC SENSORS

Robotic sensors are used in robotic system for a variety of tasks such as to detect the work piece, its existence and obstacles present in the environment.

Greater use of sensors and more intelligence should lead to a reduction of this uncertainty and because the machines can work 24 hours a day, should also lead to higher Productivity. A sensor is a type of transducer. The definition of sensor according to the Instrument Society of America is "A device which provides a usable output in response to a specified measure and". Here the output is an „electrical quantity "and measure and is a „physical quantity, property Or condition which is measured". Robotic sensor may be contacting (tactile) or non-contacting type. Contact sensor detect the change in position, acceleration, force, torque etc. at the end effecter, whereas non-contacting sensor detect presence, distance, features of work piece etc.

## Mobile Robotic Sensor

Robotics has come a long way, especially for mobile robots. In the past, mobile robots were controlled by heavy, large, and expensive computer systems that could not be carried and had to be linked via cable or wireless devices. Today, however, we can build small mobile robots with numerous actuators and sensors that are controlled by inexpensive, small, and light embedded computer systems that are carried on-board the robot. Mobile robots are designed to move from one place to another. Wheels, tracks, or legs allow the robot to traverse a terrain. Mobile robots may also feature an arm like appendage that allows them to manipulate objects. Of the two stationary or mobile the mobile robot is probably the more popular project for hobbyists to build.

## Contact or Tactile Sensors

Contact sensor uses transducer for the sensing operation. Some mostly used sensors are potentiometer, strain gauge etc. Contact or touch sensors are one of the most common sensors in robotics. These are generally used to detect a change in position, velocity, acceleration, force, or torque at the manipulator joints and the end-effecter. There are two main types, bumper and tactile. Bumper type detect whether they are touching anything, the information is either Yes or No. They cannot give information about how hard is the contact or what they are touching. Tactile sensor are more complex and provide information on how hard the sensor is touched, or what is the direction and rate of relative movement.

Tactile sensors that measure the touch pressure rely on strain gauges or pressure sensitive resistances. Variations of the pressure sensitive resistor principle include carbon fibers, conductive rubber, piezoelectric crystals, and piezodiodes. These resistances can operate in two different modes: The material itself may conduct better when placed under pressure, or the pressure may increase some area of electrical contact with the material, allowing increased current flow. Pressure sensitive resistors are usually connected in series with fixed resistances across a DC voltage supply to form a voltage divider. The fixed resistor limits the current through the circuit should the variable resistance become very small. The voltage across the pressure variable resistor is the output of the sensor and is proportional to the pressure on the resistor. The relationship is usually non-linear, except for the piezodiode, which has a linear output over a range of pressures. An analog to digital converter is necessary to interface these sensors with a computer. Different contact sensing as are:

- Touch and force sensing.

- Proximity or displacement sensing.

- Slip sensing.

## Three Axis Tactile Sensor

Three-axis tactile sensors as et al the robotic hand was composed of two robotic fingers equipped with three axis tactile sensors. Using the robotic hand, it was found that tri-axial tactile data generated the trajectory of the robotic fingers, even if a simple initial trajectory was provided for the control program. In the verification test, the robotic hand screwed a bottle cap to close it. The tactile sensor is composed of a CCD camera, an acrylic dome, a light source, and a computer as shown in Fig. The light emitted from the light source is directed into the acrylic dome. Contact phenomena are observed as image data, which are acquired by the CCD camera and transmitted to the computer to calculate the three-axis force distribution. The sensing element presented in this paper is comprised of a columnar feeler and eight conical feelers. The sensing elements, which are made of silicone rubber, are designed to maintain contact with the conical feelers and the acrylic dome and to make the columnar feelers touch an object.

When the three components of the force vector, $Fx$, $Fy$, and $Fz$, are applied to the tip of the columnar feeler, contact between the acrylic dome and the conical feelers is measured as a distribution of gray-scale values, which are transmitted to the computer. $Fx$, $Fy$, and $Fz$ values are calculated using integrated gray-scale value G and the horizontal displacement of the centric of gray-scale distribution. We are currently designing a multi-fingered robotic hand for general-purpose use in robotics. The robotic hand includes links, fingertips equipped with the three-axis tactile sensor, and micro actuators. Each micro actuator, which consists of an AC servo-motor, a harmonic drive, and an incremental encoder, is particularly developed for application to a multi-fingered hand. Since the tactile sensors must be fitted to a multi-fingered hand, we are developing a fingertip that includes a hemispherical three-axis tactile sensor.

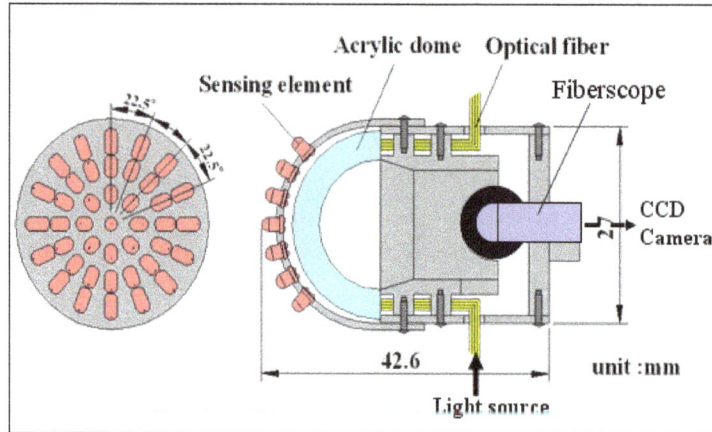

Three-Axis Tactile Sensor System.

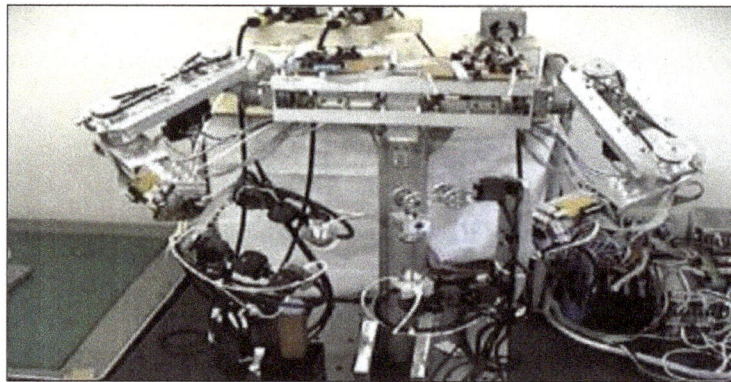

Two-hand-arm robot equipped with optical three axis tactile sensor.

## Non-Contacting Sensor

Non-contacting sensors are also a very important type of sensor, which detect parametric information about the environment of the object. It is used to detect the existence, distance and features of the object.

There are mainly six types of non-contacting sensor are as:

- Visual and optical sensor.
- Magnetic and inductive sensor.
- Capacitive sensor.
- Resistive sensor.
- Ultrasonic and sonar sensor.
- Air pressure sensor.

Visual and optical sensors operate by transforming light into an electrical signal. The photo detectors can be as simple as a single photo diode or as complex as a television camera. With stereo cameras, robotic vision systems are analogous to the human sense of sight.

Magnetic sensor is a type of non-contacting sensor which converts the magnetic energy into electrical signal. By this electrical signal it is able to determine the velocity, proximity of any metallic object.

Capacitive sensor is similar to magnetic sensor. The most common capacitive probes are flat disks or flat metal sheets. Probes are electrically isolated from their housings by guard electrodes insuring that the electric field produced is perpendicular to the sensor. Systems can make measurements in 100 microseconds with resolutions of 1/10 of a micron, and probe diameters range from thousandths of an inch to several feet.

Resistive sensing determines the distance between a robot arc welder and the welding seam. This sensing is done by means of varying resistance of the welding are. Figure shows the basic technique for through the arc position sensing.

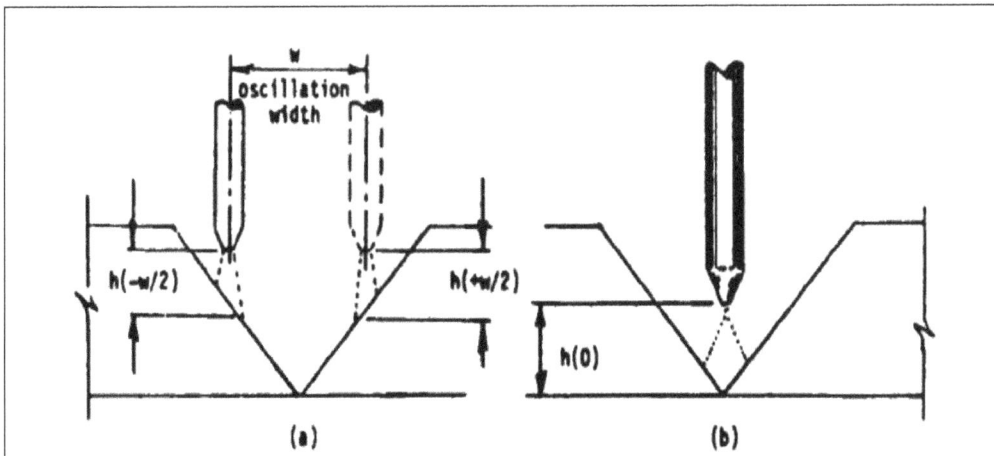

Welding Technique For Through he Arc Sensing.

Ultrasonic and sonar sensors are detected position, velocity, orientation etc. using high frequency acoustic waves.

## Light Sensors

A Light sensor is used to detect light and create a voltage difference. The two main light sensors generally used in robots are Photoresistor and Photovoltaic cells. Other kinds of light sensors like Phototubes, Phototransistors, CCD's etc. are rarely used.

Photoresistor is a type of resistor whose resistance varies with change in light intensity; more light leads to less resistance and less light leads to more resistance. These inexpensive sensors can be easily implemented in most light dependant robots.

Photovoltaic cells convert solar radiation into electrical energy. This is especially helpful if you are planning to build a solar robot. Although photovoltaic cell is considered as an energy source, an intelligent implementation combined with transistors and capacitors can convert this into a sensor.

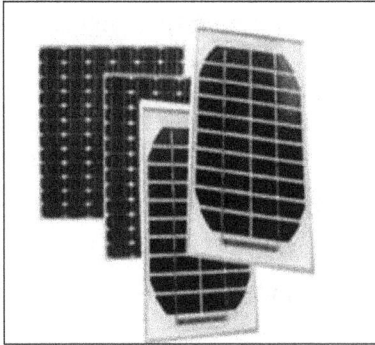
Photo voltaic cells (solar cells).

Photoresistor (LDR).

## Sound Sensor

As the name suggests, this sensor (generally a microphone) detects sound and returns a voltage proportional to the sound level. A simple robot can be designed to navigate based on the sound it receives. Imagine a robot which turns right for one clap and turns left for two claps. Complex robots can use the same microphone for speech and voice recognition.

Sound sensor (Mic).

Implementing sound sensors is not as easy as light sensors because Sound sensors generate a very small voltage difference which should be amplified to generate measurable voltage change.

## Temperature Sensor

Temperature Sensors.

What if your robot has to work in a desert and transmit ambient temperature? Simple solution is to use a temperature sensor. Tiny temperature sensor ICs provide voltage difference for a change in temperature. Few generally used temperature sensor IC's are LM34, LM35, TMP35, TMP36, and TMP37.

## Contact Sensor

Contact sensors are those which require physical contact against other objects to trigger. A push button switch, limit switch or tactile bumper switch are all examples of contact sensors. Limit Switch These sensors are mostly used for obstacle avoidance robots. When these switches hit an obstacle, it triggers the robot to do a task, which can be reversing, turning, switching on a LED, Stopping etc. There are also capacitive contact sensors which react only to human touch. Touch screen Smart phones available these days use capacitive touch sensors. Contact Sensors can be easily implemented, but the drawback is that they require physical contact. In other words, your robot will not turn until it hits an object. A better alternative is to use a proximity sensor.

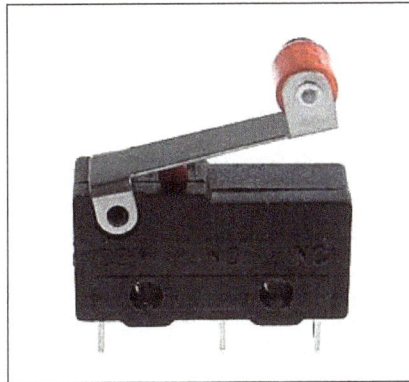

Limit switch (contact sensor).

## Proximity Sensor

This is a type of sensor which can detect the presence of a nearby object within a given distance, without any physical contact. The working principle of a Proximity sensor is simple. A transmitter transmits an electromagnetic radiation or creates an electrostatic field and a receiver receives and analyzes the return signal for interruptions. There are different types of Proximity sensors and we will discuss only a few of them which are generally used in robots.

1.  Infrared (IR) Transceivers: An IR LED transmits a beam of IR light and if it finds an obstacle, the light is simply reflected back which is captured by an IR receiver. Few IR transceivers can also be used for distance measurement.

2.  Ultrasonic Sensor: These sensors generate high frequency sound waves; the received echo suggests an object interruption. Ultrasonic Sensors can also be used for distance measurement.

3.  Photoresistor: Photoresistor is a light sensor; but, it can still be used as a proximity sensor. When an object comes in close proximity to the sensor, the amount of light changes which in turn changes the resistance of the Photoresistor. This change can be detected and processed.

There are many different kinds of proximity sensors and only a few of them are generally preferred for robots. For example, Capacitive Proximity sensors are available which detects change in capacitance around it. Inductive proximity sensor detects objects and distance through the use of induced magnetic field.

## Distance Sensor

Most proximity sensors can also be used as distance sensors, or commonly known as Range Sensors; IR transceivers and Ultrasonic Sensors are best suited for distance measurement:

1. Ultrasonic Distance Sensors: The sensor emits an ultrasonic pulse and is captured by a receiver. Since the speed of sound is almost constant in air, which is 344m/s, the time between send and receive is calculated to give the distance between your robot and the obstacle. Ultrasonic distance sensors are especially useful for underwater robots.

2. Infrared Distance sensor: IR circuits are designed on triangulation principle for distance measurement. A transmitter sends a pulse of IR signals which is detected by the receiver if there is an obstacle and based on the angle the signal is received, distance is calculated. SHARP has a family of IR transceivers which are very useful for distance measurement. A simple transmit and receive using a couple of transmitters and receivers will still do the job of distance measurement, but if you require precision, then prefer the triangulation method.

3. Laser range Sensor: Laser light is transmitted and the reflected light is captured and analyzed. Distance is measured by calculating the speed of light and time taken for the light to reflect back to the receiver. These sensors are very useful for longer distances.

4. Encoders: These sensors (not actually sensors, but a combination of different components) convert angular position of a shaft or wheel into an analog or digital code. The most popular encoder is an optical encoder which includes a rotational disk, light source and a light detector (generally an IR transmitter and IR receiver). The rotational disk has transparent and opaque pattern (or just black and white pattern) painted or printed over it. When the disk rotates along with the wheel the emitted light is interrupted generating a signal output. The number of times the interruption happens and the diameter of the wheel can together give the distance travelled by the robot.

5. Stereo Camera: Two cameras placed adjacent to each other can provide depth information using its stereo vision. Processing the data received from a camera is difficult for a robot with minimal processing power and memory. If opted for, they make a valuable addition to your robot.

There are other stretch and bend sensors which are also capable of measuring distance. But, their range is so limited that they are almost useless for mobile robots.

## Pressure Sensors

As the name suggests, pressure sensor measures pressure. Tactile pressure sensors are useful in robotics as they are sensitive to touch, force and pressure. If you design a robot hand and need to measure the amount of grip and pressure required to hold an object, then this is what you would want to use.

## Tilt Sensors

Tilt sensors measure tilt of an object. In a typical analog tilt sensor, a small amount of mercury is suspended in a glass bulb. When mercury flows towards one end, it closes a switch which suggests a tilt.

## Navigation/Positioning Sensors

The name says it all. Positioning sensors are used to approximate the position of a robot, some for indoor positioning and few others for outdoor positioning.

- GPS (Global Positioning System): The most commonly used positioning sensor is a GPS. Satellites orbiting our earth transmit signals and a receiver on a robot acquires these signals and processes it. The processed information can be used to determine the approximate position and velocity of a robot. These GPS systems are extremely helpful for outdoor robots, but fail indoors. They are also bit expensive at the moment and if their prices fall, very soon you would see most robots with a GPS module attached.

- Digital Magnetic Compass: Similar to a handheld magnetic compass, Digital Magnetic compass provides directional measurements using the earth's magnetic field which guides your robot in the right direction to reach its destination. These sensors are cheap compared to GPS modules, but a compass works best along with a GPS module if you require both positional feedback and navigation. Philips KMZ51 is sensitive enough to detect earth's magnetic field.

- Localization: Localization refers to the task of automatically determining the location of a robot in complex environment. Localization is based on external elements called landmarks which can be either artificially placed landmarks, or natural landmark. In the first approach, artificial landmarks or beacons are placed around the robot, and a robot's sensor captures these signals to determine its exact location. Natural landmarks can be doors, windows, walls, etc. which are sensed by a robots sensor / vision system (Camera). Localization can be achieved using beacons which generate Wi-Fi, Bluetooth, Ultrasound, Infrared, Radio transmissions, Visible Light, or any similar signal.

## Acceleration Sensor

An accelerometer is a device which measures acceleration and tilt. There are two kinds of forces which can affect an accelerometer: Static force and Dynamic Force.

- Static Force: Static force is the frictional force between any two objects. For example earth's gravitational force is static which pulls an object towards it. Measuring this gravitational force can tell you how much your robot is tilting. This measurement is exceptionally useful in a balancing robot, or to tell you if your robot is driving uphill or on a flat surface.

- Dynamic force: Dynamic force is the amount of acceleration required to move an object. Measuring this dynamic force using an accelerometer tells you the velocity/speed at which your robot is moving. We can also measure vibration of a robot using an accelerometer, if in any case you need to.

Accelerometer comes in different flavors. Always select the one which is most appropriate for your robot. Some of the factors which you need to consider before selecting an accelerometer are:

- Output Type: Analog or Digital.

- Number of Axis: 1,2 or 3.

- Accelerometer Swing: ±1.5g, ±2g, ±4g, ±8g, ±16g.

- Sensitivity: Higher or Lower (Higher the better).

- Bandwidth.

## Gyroscope

A gyroscope or simply Gyro is a device which measures and helps maintain orientation using the principle of angular momentum. In other words, a Gyro is used to measure the rate of rotation around a particular axis. Gyroscope is especially useful when you want your robot to not depend on earth's gravity for maintaining Orientation.

## IMU

Inertial Measurement Units combine properties of two or more sensors such as Accelerometer, Gyro, Magnetometer, etc, to measure orientation, velocity and gravitational forces. In simple words, IMU's are capable of providing feedback by detecting changes in an objects orientation (pitch, roll and yaw), velocity and gravitational forces. Few IMUs go a step further and combine a GPS device providing positional feedback.

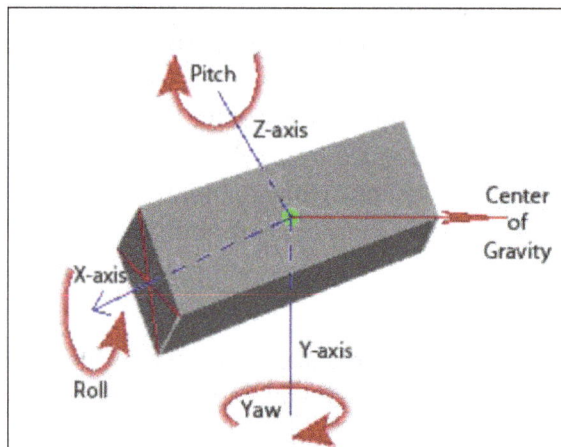
Pitch, Roll and Yaw of an object.

## Voltage Sensors

Voltage sensors typically convert lower voltages to higher voltages, or vice versa. One example is a general Operational-Amplifier (Op-Amp) which accepts a low voltage, amplifies it, and generates a higher voltage output. Few voltage sensors are used to find the potential difference between two ends (Voltage Comparator). Even a simple LED can act as a voltage sensor which can detect a voltage difference and light up.

## Current Sensors

Current sensors are electronic circuits which monitor the current flow in a circuit and output either a proportional voltage or a current. Most current sensors output an analog voltage between 0V to 5V which can be processed further using a microcontroller.

## Other Sensors for Robots

There are hundreds of sensors made today to sense virtually anything you can think of, and it is almost impossible to list all available sensors. Apart from those mentioned above, there are many other sensors used for specific applications. For example: Humidity Sensors measures Humidity; Gas sensors are designed to detect particular gases (helpful for robots which detects gas leaks); Potentiometers are so versatile that they can be used in numerous different applications; Magnetic Field Sensors detect the strength of magnetic field around it.

## Types of Industrial Robot Sensors

In the industrial and automation world, machines need sensors to provide them with the required information to execute a proper operation. A lot of sensors can be added to different robots to increase their adaptability. Industrie see a lot of collaborative robots having integrated force torque sensors and cameras in order to have a better perspective on their operations and also to provide the safest workspace.

## 2D Vision

2D is vision is basically a video camera that can perform a lot of different things. From detecting movement to localization of a part on a conveyor. 2D vision has been on the market for a long time and is here to stay. Many smart cameras out there can detect parts and coordinate the part position for the robot so that it can then adapt its actions to the information it receives.

## 3D Vision

3D vision is much more recent phenomenon as compared to 2D vision. A tri-dimensional vision system has to have 2 cameras at different angles or use laser scanners. This way, the third dimension of the object can be detected. Once again many applications use 3D vision. Bin picking, for example, can use 3D vision to detect objects in a bin and recreate the part in 3D, analyze it and pick it the best way possible.

## Force Torque Sensor

While vision gives eyes to the robot, force torque sensors give touch to the robot wrist. Here the robot uses a force torque sensor (FT sensor) to know the force that the robot is applying with its end of arm tooling. Most of the time, the FT sensor is located between the robot and the tool. This way, all the forces that are applied on the tool are monitored.

Applications such as assembly, hand-guiding, teaching and force limitation can be done with this device. Industries have developed a technology called Kinetiq Teaching that allows industrial welding robots to be taught via hand-guiding. All of this using a force torque sensor to monitor the motions of the "teacher". Industries also have a FT Sensor that is sold separately to accomplish any number of applications that might require force sensing.

## Collision Detection Sensor

This kind of sensor can have different forms. As the main applications of these sensors is to provide a safe working environment for human workers, the collaborative robots are most likely to use them. Some sensors can be some kind of tactile recognition systems, where if a pressure is sensed on a soft surface, a signal will be sent to the robot to limit or stop its motions.

You can also see this kind of sensor directly built into the robot. Some companies use accelerometers, some use current feedback. In either case, when an abnormal force is sensed by the robot the emergency stop is released. This provides a safer environment. The safest environment is an environment with no risk of collision.

## Safety Sensors

With the introduction of industrial robots in collaborative mode, industry has to react with a way to protect its workers. These sensors can really appear in a lot of different shapes. From cameras to lasers, a safety sensor is designed to tell the robot that there is a presence around it. Some safety systems are configured to slow down the robot once the worker is in a certain area/space and to stop it once the worker is too close.

A simple example of safety sensors would be the laser on your garage door. If the laser detects an obstacle, the door immediately stops and goes backwards to avoid a collision. This can be a good comparision to what safety sensors are like in the robotic industry.

## Part Detection Sensors

For applications that require you to pick parts, you probably have no clue if the part is in the gripper or if you just missed it (assuming you don't have a vision system yet!). Well, a part detection application gives you feedback on your gripper position. For example, if a gripper misses a part in its grasping operation, the system will detect an error and will repeat the operation again to make sure the part is well grasped.

Adaptive Grippers have part detection systems that don't need any sensors. In fact, Grippers are designed to grasp parts with a given force. So the Gripper doesn't need to know that the part is there or not, it will only apply enough force to get the best grip on the object. Once the required force is reached, you know that the object is in the Gripper and that it is ready for the next step in the operation.

## Others

Of course there are a lot of other sensors that can be fitted to your robotic cell that are very specific to your application. Sensors that are capable to do seam tracking in welding operations are a good example where a specific sensor is necessary.

Tactile sensors are also becoming more popular these days. This kind of sensor is, most of the time, fitted on a gripper to detect and feel what is in it. Sensors are usually able to detect forces and draw an array of vectors with the force distribution. This shows the exact position of an object and allows you to control the position and the grasping force of the end effector. Some tactile sensors can also detect heat variation.

Finally, sensors are key components to leveraging software intelligence. Without such sensors, advanced operations wouldn't be possible. They bring a lot of complexity to the operation, but they also insure good control during the process.

## References

- 8-main-components-of-robots, education: preservearticles.com, Retrieved 10 February, 2019

- Trejos AL, et al (Sep 2009). "Robot-assisted Tactile Sensing for Minimally Invasive Tumor Localization." International Journal of Robotics Research 28 (9): 1118-1133

- Actuators, robotics-technology, robotics: electronicsteacher.com, Retrieved 11 March, 2019

- Pinto, Jim (October 1, 2003). "Fully automated factories approach reality". Automation World. Retrieved December 3,2018

- Robotic-arm-and-their-application: meee-services.com, Retrieved 12 April, 2019

- Zeghloul, Saïd; Laribi, Med Amine; Gazeau, Jean-Pierre (21 September 2015). Robotics and Mechatronics: Proceedings of the 4th iftomm International Symposium on Robotics and Mechatronics. Springer. ISBN 9783319223681 – via Google Books

- Robotic-manipulation, research: northwestern.edu, Retrieved 13 May, 2019

- Žlajpah, Leon (2008-12-15). "Simulation in robotics". Mathematics and Computers in Simulation. 79 (4): 879–897. Doi:10.1016/j.matcom.2008.02.017

- Types-of-robot-sensors, sensors, knowledge: robotplatform.com, Retrieved 14 June, 2019

- Zunt, Dominik. "Who did actually invent the word "robot" and what does it mean?". The Karel Čapek website. Archived from the original on 2013-01-23. Retrieved 2017-02-05

- 7-Types-of-Industrial-Robot-Sensors, bid: robotiq.com, Retrieved 15 July, 2019

# 4

# Locomotion and Control

The collection of methods that are used by robots that enable them to transport themselves from place to place is known as robotics locomotion, legged locomotion, wheeled locomotion, bio-inspired robotic locomotion, etc. are some concepts that fall under its domain. This chapter closely examines these concepts of robotics locomotion and control to provide an extensive understanding of the subject.

## LOCOMOTION

Locomotion is directional movement that enables someone or something to move from one location to another. The word derives from the Latin words locō (place) and mōtiō (to move).

The study of locomotion informs many areas of science, medicine and technology. The mechanisms of locomotion may be applied in biomimetics (biomimicry), the development of human-made processes, substances, devices or systems that imitate nature. In robotics, for example, designers imitate human movement to create life-like androids.

Locomotion is also an important area of endeavor in video game art and design and virtual reality (VR). Creating realistic locomotion for digital content requires an understanding of how that movement is accomplished and what it looks like in the physical world. In VR gaming, locomotion usually refers to systems that allow the user to navigate through the virtual environment.

## ROBOTIC LOCOMOTION

The mechanism that makes a robot capable of moving in its environment is called as robot locomotion.

### Legged Locomotion

A legged robot is well suited for rough terrain; it is able to climb steps, to cross gaps which are as large as its stride and to walk on extremely rough terrain where, due to ground irregularities, the use of wheels would not be feasible. To make a legged robot mobile each leg must have at least two degrees of freedom (DOF). For each DOF one joint is needed, which is usually powered by one servo. Because of this a four legged robot needs at least eight servos to travel around. Figure shows the energy consumption of different locomotion concepts. It strikes that the power consumption of legged locomotion is nearly two orders of magnitude more inefficient than of wheeled locomotion on hard, flat surface (e.g. railway wheel on steel). One reason for this is that wheeled locomotion requires in general fewer motors than legged locomotion.

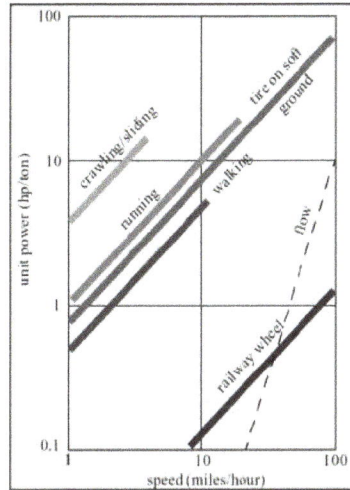

Power consumption of several locomotion mechanisms.

When the surface becomes soft wheeled locomotion offers some inefficiency, due to increasing rolling friction more motor power is required to move. As figure shows legged locomotion is more power efficient on soft ground than wheeled locomotion, because legged locomotion consists only of point contacts with the ground and the leg is moved through the air. This means that only a single set of point contacts is required, so the quality of the ground does not matter, as long as the robot is able to handle the ground. But exactly the single set of point contacts offers one of the most complex problem in legged locomotion, the stability problem.

## Stability

Stability is of course a very important issue of a robot, because it should not overturn. Stability can be divided into the static and dynamic stability criterion.

Example of a stool.

Static stability means that the robot is stable, with no need of motion at every moment of time. Static stability is explained by an easy example: Figure shows a stool with three legs. Balance is maintained as long as the centre of mass is completely within the red triangle, which is set by the stools' footprints. This triangle is called support polygon. The support polygon is the convex hull which is set by the ground contact points. Of course, in case of more ground contact points, the polygon can be a quadrangle or a pentagon or a different geometrical figure. More in general the following must hold to support static stability: Static stability is given, when the centre of mass is completely within the support polygon and the polygon's area is greater than zero, therefore static stability requires at least three points of ground contact.

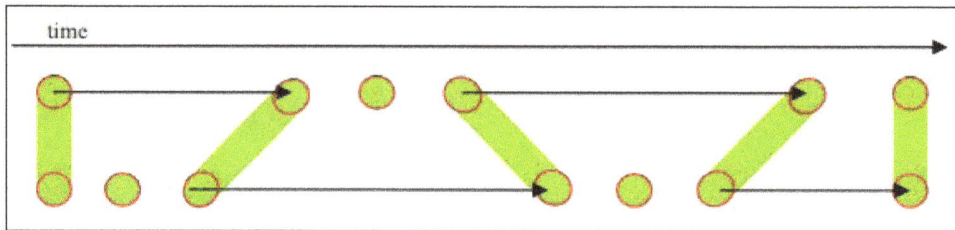

A bipedal robots' gait cicle; the red circles indicate the footprints; the green area is the supporting area; the leg movement is expressed by the arrows.

To achieve statically stable walking a robot must have a minimum number of four legs, because during walking at least one leg is in the air. Statically stable walking means that all robots' motion can be stopped at every moment in the gait cycle without overturning. Most robots which are able to walk static stable have six legs, because walking static stable with four legs means that just one leg can be lifted at the same time (lifting more legs will reduce the support polygon to a line), so walking becomes slowly Most two legged walking machines are dynamically stable for several reasons. Human like robots have relatively small footprints, because of this the support polygon is almost a line (in the double support phase, when both foots are connected with the ground) which is even reduced to a single point (in the single support phase, when just one foot has ground contact) during walking. Therefore the robot must actively balance itself to prevent overturning. Figure b) shows the changing support polygon in a bipedal walking machines' gait cycle. In face of that the robots' centre of mass has to be shifted actively between the footprints. But the robots exact centre of mass is hard to predict due to the high dynamic of walking (for example because of the force which is imparted to whole robot when one leg swings forward).

The realization of bipedal dynamic stable walking machines is due to the continuous danger of overbalance a high complex problem for engineers, which is just solved for some special cases.

## Leg Configuration

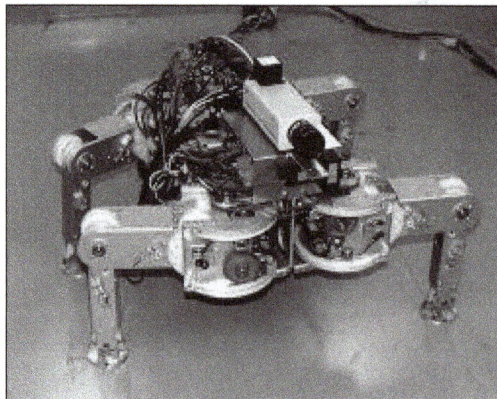

Leg of Titan VIII.

To move a leg forward at least two degrees of freedom are required, one for lifting and one for swinging. Most legs have three degrees of freedom; this makes the robot able to travel in rougher terrain and to do more complex manoeuvres. Figure shows the leg of the Titan VIII robot from the Tokyo Institute of Technology. This leg has three degrees of freedom. In general, adding degrees of freedom to a robots leg means increasing the manoeuvrability of the robot, the range of terrain on which it can travel and the ability to travel in a variety of gaits.

But adding degrees of freedom causes also some disadvantages, because for moving additional joints and more servos are required, this increases the power consumption and the weight of the robot. Furthermore controlling the robot becomes more complex, because more motors have to be controlled and actuated at the same time.

If the robot has more than one leg there is the issue of leg coordination for locomotion. The total number of possible gaits in which a robot can travel depends on the number on legs it has. The gait is a periodic sequence of lift and release events for each leg. If a robot has k legs the number of possible events N is, accordant to,

$$N=(2k-1)!$$

In case of a bipedal walking machine (k=2) the number of possible events is $N=(2k-1)! = (2*2-1)! = 3! = 6$

So there are six possible different events, these are:

1. Lift left leg.

2. Release left leg.

3. Lift right leg.

4. Release right leg.

5. Lift both legs together.

6. Release both legs together.

In case of k=6 legs there are already 39916800 possible events, in face of that, controlling a six legged robot is because of the large number of possible events more complex than controlling a two legged robot. But robots with fewer legs have some other problems; one of the most complex problems is stability as mentioned before. In the following different leg configurations, advantages/disadvantages of these and examples of robots are shown.

## One Leg

One leg is of course the minimum number of legs which a legged robot can have. A smaller number of legs reduces body mass of the robot and no leg coordination is needed. One-legged locomotion requires just a single point of ground contact; this makes the robot amenable to travel the roughest terrain. As an example the robot is able to overcome an obstacle like a gap that is larger than its stride by talking a running start. A multi legged robot that can not run is just able to cross gaps that are as large as its reach. But the single point of ground contact offers the main problem for single legged robots – stability. Static stability is impossible even when the robot is stationary, because the support polygon is reduced to a single point. So singled legged robots must be dynamically stable, that means that the robot has to actively balance itself either by changing its centre of gravity or by imparting corrective forces. One of the first successful one-legged robots was the one leg hopper developed by Marc Raibert in 1983.

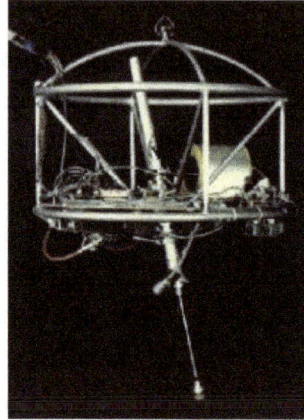

One leg hopper.

Raibert's hopper is not able to be stable when it is stationary, so it has to hop all the time. To support locomotion and stability there is of course the need of controlling the robot. Raiberts' hopper uses a simple controller, which divides the control problem into three independent parts. These parts are hopping height, velocity and attitude.

- Hopping height: The control system controls hopping height by manipulating hopping energy. The leg is springy, so hopping is a bouncing motion that is generated by an actuator (an external air-pressure pump) that excites the leg. Hopping height is determined by the energy recovered from the previous hop, the losses in the hopping cycle and thrust developed in the actuator. Height is regulated by adjusting the amount of thrust on each cycle to just make up for losses.

- Velocity: The control system manipulates forward velocity by placing the foot with respect to the centre of the CG-print on each step. The CG-print is the locus of points on the ground over which the centre of gravity of the system will pass during stance. Displacing the foot from the centre of the CG-print causes the system to run either faster or slower. The control system calculates the length of the CG-print from the measured forward velocity and an estimate of the duration of stance. The error in forward velocity determines a foot position that will maintain the correct speed of forward travel.

- Attitude: The control system maintains an erect body posture during running, by generating hip torques during stance that servo the body angle. During stance friction between the foot and ground permits large torques to be applied to the body without causing large accelerations of the leg. These torques are used to implement a simple proportional servo that moves the body toward an erect posture once each step.

## Two Legs

Bipedal walking robots have become very popular in the last ten years; two of the most well known examples are QRIO from Sony and Asimo from Honda. Qrio has a weight of 7 kg and a height of

58 cm, each leg has six degrees of freedom; Asimo has a weight of 210 kg, a height of 1.82 cm and a maximum walking speed of 2 km/h, each leg has six degrees of freedom.

QRIO (Sony).                Asimo (Honda).

Two legged robots are already able to walk, run, jump, dance and travel up and down stairs, but stability is still a problem for bipedal robots, because they have to be dynamically stable. There is no general algorithm to solve the problem of dynamic stability for bipedal robots; often used approaches are based on the zero moment point (ZMP). Examples of robots using this approach are QRIO and Asimo. The rudiment idea of this approach is to maintain balance by planning footprint positioning. The ZMP is the point where the robot has to base on to keep its balance. When the robot should move forward it has first to compute the ZMP and after that it has to step the appropriate leg exactly to the computed position. The Zero Moment Point1 (ZMP) is often described in robotics as the point on the ground where all momentums are equal to zero. The ZMP can be computed with equation $x_{ZMP} = \dfrac{\sum_i m_i(z+g)x_i - \sum_i m_i x z_i - \sum_i I_{iy}\theta_{iy}}{\sum_i m_i(z+g)}$ and

$$y_{ZMP} = \frac{\sum_i m_i(z+g)y_i - \sum_i m_i y z_i - \sum_i I_{ix}\theta_{ix}}{\sum_i m_i(z+g)} \text{ (accordant to):}$$

$$x_{ZMP} = \frac{\sum_i m_i(z+g)x_i - \sum_i m_i x z_i - \sum_i I_{iy}\theta_{iy}}{\sum_i m_i(z+g)}$$

$$y_{ZMP} = \frac{\sum_i m_i(z+g)y_i - \sum_i m_i y z_i - \sum_i I_{ix}\theta_{ix}}{\sum_i m_i(z+g)}$$

Where $(x_{ZMP},\ _{ZMP}, 0)$ are the ZMP coordinates in the Cartesian coordinate system, $(x_i, y_i, z_i)$ is the mass centre of the link $i$, $m_i$ is the mass of the link $I$, and g is the gravitational acceleration.

$I_x$ and $I_y$ are the inertia moment components, $\theta_{ix}$ and $\theta_{iy}$ are the angular velocity around the axes x and y (taked as a point from the mass centre of the link i). Figure shows an example of a robot's mass distribution.

To support stability the ZMP has to be completely within the support polygon. If one leg is in the air, the support polygon is equal to the shape of the foot which is connected with the ground, so the ZMP has to be completely within the footprint to support stability. If both feet are connected with the ground the ZMP can be within the area which is built by the two footprints.

Asimo and QRIO are robots which are already able to walk freely through their environment, but they have to take care of their balance at every time. Spring Flamingo is a robot which was developed in 1996 to 2000 by Jerry Pratt. Each leg of Spring Flamingo, which is inspired by a flamingo, has three degrees of freedom, these are realized by a hip, knee and ankle joint.

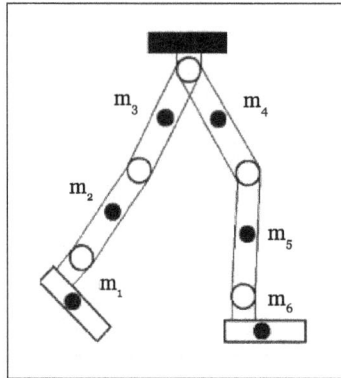

Example of a robot's mass distribution.

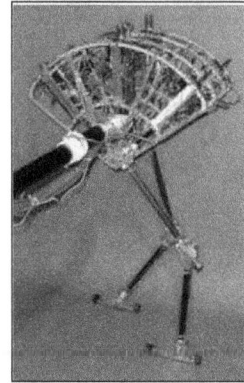

Spring Flamingo (MIT).

In addition, each leg has a kneecap that limits the knee joint angle. Spring Flamingo's research goal is not to solve the stability problem, because it is helt by a bar at every moment of movement, as figure shows, so it is not able to overturn. Due to this the stability problem is suppressed and the focus of research is on developing various walking algorithms, motion description and control techniques and force control actuation techniques.

One benefit of a two legged locomotion is that the total weight of the robot is reduced due to fewer legs (a six legged robot has much more leg mass and because of this more body mass), but this advantage creates another problem. Each leg must have sufficient capacity to support the full weight of the robot, in case of four or six legged robots the weight of the robot's body is distributed to more legs.

An important feature of bipedal robots is their anthropomorphic shape, they can be build in human like dimensions, which makes them predestinated for research in human robot interaction.

## Four Legs

One of the most famous four legged robot is Sony's Aibo . Some of Aibo's most interesting features are a stereo microphone, which enables it to pick up surrounding sounds, a head sensor to notice a person who tabs its head, eye lights (these light up in blue, green or red) to indicate Aibo's emotional state, a colour camera to search for objects and recognize them by colour and movement, and speakers to emit sounds. Some four legged robots are also well adapted for research in human robot interaction, if they have an animal shape (like Aibo). Humans can treat them as a pet and might develop an emotional relationship to them.

Another example of a quadruped robot is Titan VIII, which was developed at Tokyo Institute of Technology. Titan VIII has a weight of 9 kg, a height of 0.25 m and each leg has six degrees of freedom.

Most four legged robots use dynamic stable walking (like nearly all four legged animals), because static stable walking requires at least three points of ground contact. This means that just one leg can be lifted at the same time and so walking becomes slowly; in case of dynamic stability the number of ground contact points can vary from zero, when the robot is jumping, to the total number of legs, when the robot is stationary. One possible dynamic stable gait of Titan VIII is a trot gait, where the two diagonal legs are lifted at the same time.

Aibo (Sony).

Titan VIIIs' dynamic stable walking is also based on the ZMP. To support stability during walking the ZMP has to be about the diagonal line, which is set by the two legs with ground contact. More information about realization of this concept can be taken from.

Titan VIII (Tokyo Institute of Technology).

As mentioned before four legged robots are able to walk statically stable. Figure a) shows a simplified model (2D+1 Model) of a four legged robot. The body of the robot is described by a polygon, the legs by straight lines and the footprints by empty or filled circles (empty if the leg is raised, filled if the leg has ground contact). The centre of the polygon is considered as the robots centre of mass, indicated the green point (this is of course a simplified assumption). The number one to five indicates a predefined set of leg position. That means that each leg can take one of these positions and it can be raised or set down.

The total number of possible robot configurations is equal to $5^4 \times 2^4 = 10000$, but not all of these configurations are stable. Figure b) shows an example of a stable configuration and Figure c) shows an unstable one (the red triangle is the support polygon).

One way to find static stable gaits can be done by first reducing the total number of configurations, by eliminating all those configurations which are unstable (for example by testing all configurations). After that a search in the set formed by the stable robot configuration can be done, to find sequences of configurations, which can be used as gaits.

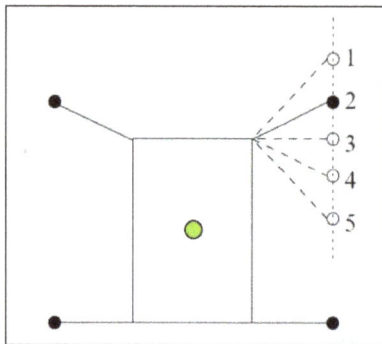

S2D+1 Model of a four Legged robot.      Stable configuration.      Unstable configuration.

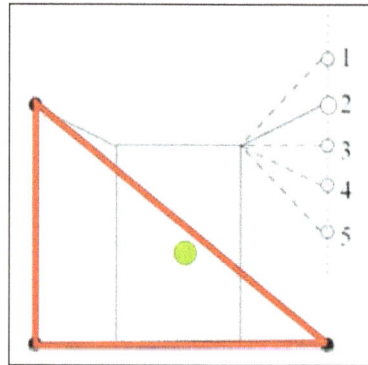

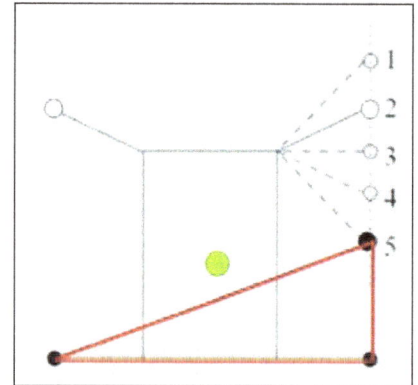

## Six Legs

Six legged locomotion is the most popular legged locomotion concept because of the ability of static stable walking. The most used static stable gait is the tripod gait, where each times the two exterior legs on the one side and the inner leg of the other side are moved together. Due to the possibility of static stable gaits the control complexity is reduced on the one hand, because there is no issue of stability control in general, but on the other hand most six legged robots legs have three degrees of freedom and six legs have to be controlled, so leg coordination becomes more complex. Six legged robots are often inspired by nature, two examples of such robots are Lauron and Genghis. One approach to reduce the complexity of controlling is to decentralize the control of the robot. Studies of the nervous system of six legged animals have shown that most six legged animals control their leg movement locally and not by brain. An often used example for six legged robots is the stick insect. Lauron III is an example of a six legged robot which is inspired it. The Lauron project began in 1993, Lauron III is the current robot which was developed in 2001and is still advanced by the Forschungszentrum Informatik, group IDS in Karlsruhe. Lauron III has a lenth of 0.5 meter, a height of 0.3 meter, width of 0.8 meter, a weight of 18 kg and each leg has three degrees of freedom. Each leg is controlled by one Siemens 80C176 microcontroller, all legs are connected among each other and with an onboard PC/104 (equipped with a Pentium II 400 and a Real Time Linux) by a CAN-Bus.

The control software of Lauron III is built (like the hardware architecture) hierarchically. The software is divided into modules for different subtask and is distributed to the microcontrollers and the PC/104. These different modules are a joint controller for each joint (gets as input an angle and sets the joint accordant to the angle), a leg controller for each leg (routes angles to the joint controllers to set the footprint to a defined position) and a gait controller (coordinates the legs). Dividing the complex overall control system into smaller subsystems makes developing the overall system easier and more understandable. Another advantage is that subsystems can be developed and tested independently without changing the whole system.

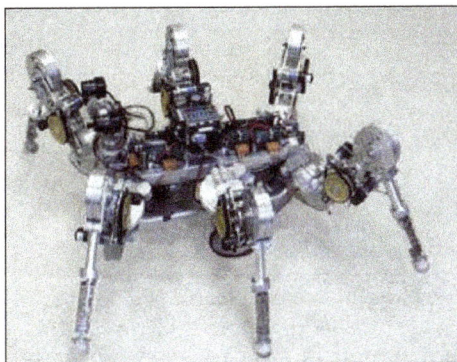

Lauron ( Forschungszentrum Informatik, group IDS in Karlsruhe).

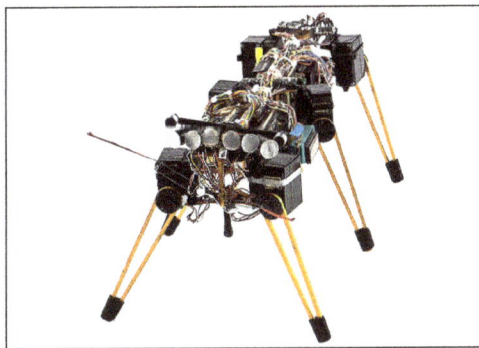

Genghis (MIT).

The problem of "how to make a legged robot walking" can be solved by programming the robot as finite automata, where all walking actions are defined before walking through the environment. The main disadvantage of this approach is that the robot has to be programmed new, when the environment changes, so the robot is not able to walk through a dynamic changing environment. Another approach is the use of reinforcement learning algorithms, by mean of these the robot is able to learn walking by it self. The object of reinforcement learning is to learn something by trial and error interaction with a dynamic environment. The subject who wants to learn something is called the agent (in this case it is of course the robot). In the standard reinforcement learning model the agent is connected to the environment via perception and interaction. The agent receives at discrete points of time $t = 0,1,2\ldots$ the state of the environment $st$ as input. Then the agent chooses an action $at$ of possible set of actions $A(s_t)$ as output and the environment changes into a new state $s_{t+1}$. To optimize the agents behaviour, the agent gets a reinforcement signal $r_{t+1} \in R$ after each action. R is the set of reinforcement signals, for example $R = \{0,1\}$, where 0 is a penalty and 1 a reward. These signals indicate the agent, when it comes into a known situation, if the decision taken the last time was good or not, so it can take the same action as the last time or it can try a different one. The goal of reinforcement learning algorithms is to choose actions that increase the long-run sum of values of the reinforcement signals. In this way the agent will learn to act in the environment.

In case of a legged robot the robot is the agent, the environment is the environment where the robot should walk through and input (including reinforcement signals) is committed with sensor (e.g. camera, touch sensor, etc). Reward can be given when the robot has moved and penalty can be given when the robot has crashed.

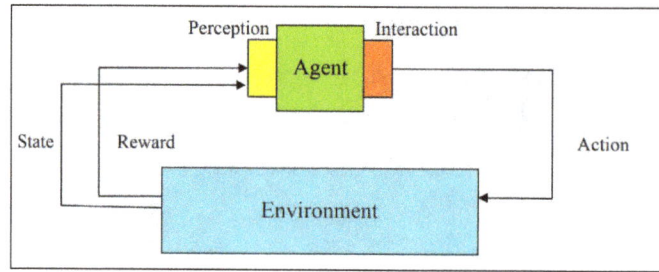
Reinforcement learning model.

An example of a six legged robot which is able to learn walking with a reinforcement learning algorithm is Genghis, developed by Rodney A. Brooks. The learning algorithm was developed by Pattie Maes and Rodney A. Brooks in 1990. Genghis is able to learn coordinating its legs to move forward. Each leg of Genghis has two elemental behaviours, swing-leg-forward and swingleg-backward (that means that the elementary leg movement is given and has not to be learned); learning means in this case that the right behaviour has to become active in the right moment to move forward. To achieve this aim a statistical method is used, that calculates in which state and how often an action ends in a positive (when Genghis moves forward) or negative result (when Genghis crash). Negative feedback is committed by two touch sensors at the bottom of Genghis; these are activated, when one or both sensors have ground contact, in case when Genghis overturns. Positive feedback is committed every time when a trailing wheel behind Genghis measures forward movement. In this way Ghenghis is able to adopt a static stable gait like the tripod gait.

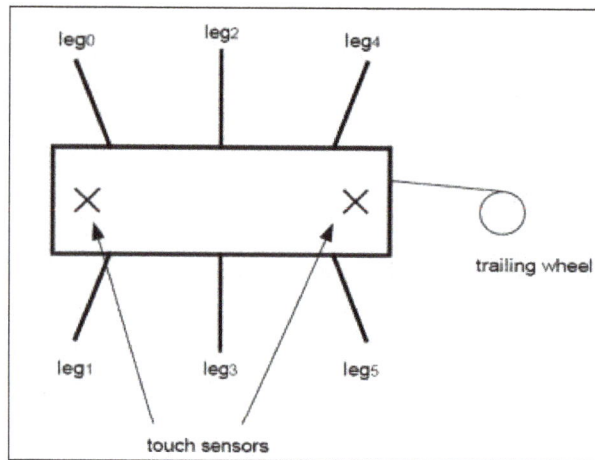
Schematic representation of Genghis' sensors.

## Wheeled Locomotion

The most popular locomotion mechanism in man made vehicles is wheeled locomotion; so it is not surprising that it is often used in mobile robotics. Reasons for this are the easy mechanical implementation of the wheel, there is no need of balance control if the vehicle has at least three or in some case two wheels and wheeled locomotion is relatively power efficient, even at high speed, as figure shows. The problems of wheeled robots are different from the problems of legged robots, as mentioned before, stability is not such a profoundly problem like it is in legged locomotion, but there are some others. The focus of research in wheeled robotics is on traction and stability in rough terrain, manoeuvrability and control.

## Wheeled Locomotion Principles

There are different types of wheeled locomotion systems. They are listed below. Each locomotion system is unique, has some advantages and disadvantages. We will discuss them one by one:

- Differential Drive.

- Car Type Drive.

- Skid Steer Drive.

- Articulated Drive.

- Synchronous Drive.

- Pivot Drive.

- Dual Differential Drive.

## Differential Drive

This is the most popular and widely used type of drive for wheeled robots, because it is the simplest and easiest to implement. There are two motors, each having an independent motion. In the first two diagrams shown above, both the motors are rotated in same direction of motion and thus the robot moves either forward or backward. In the last two diagrams, the two motors rotate such that they oppose each other's motion, thus generating a couple and creating a turning effect.

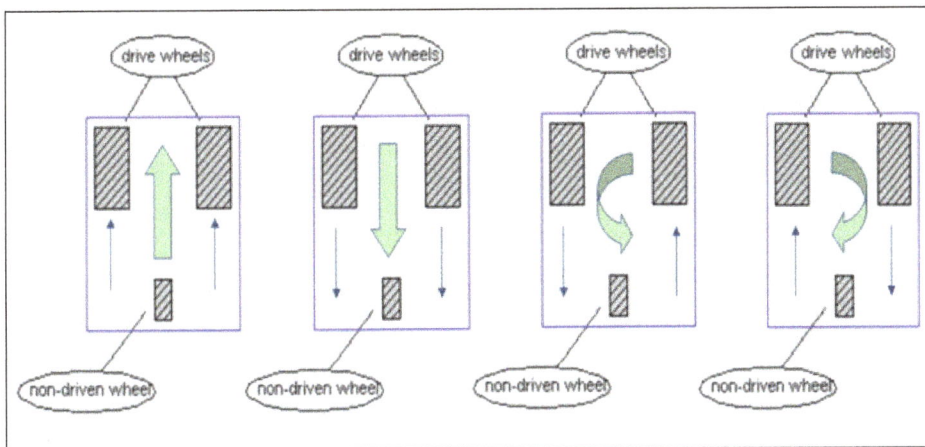

Differential Drive.

## Advantages

- Simple and easy to implement.

- Arbitrary motion can be achieved.

- In-place rotation (zero radius) can be done.

## Disadvantages

- Difficult to maintain straight line due to independent motors.

## Car Type Drive

This is the type of drive most common in real world but not in the robotic world. Here, we have a pair of wheels which direct where the robot should move, whereas movement is brought about by a different set of wheels. Translatory motion is provided by the rear wheels whereas rotational motion is provided by the front wheels. Though both motions are independent, but their interlinking results in greater accuracy.

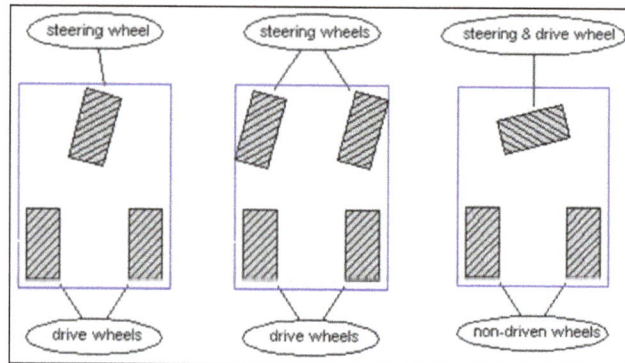

Car Type Drive.

## Advantages

- Replicates the real world.

## Disadvantages

- Difficult path planning.

- Inaccurate movement.

- Slight inaccuracy results in huge errors.

- No direct directional actuators available.

## Skid Steer Drive

It is a close relative of the differential drive system. Here, all the motors of one side are tied together as one to increase traction (e.g. tanks). Only the center motors are connected. The remaining motors move due to the force of the central motors. During turning, the wheels skid/slip over the surface. Turning can occur due to difference in the motion of the two motors.

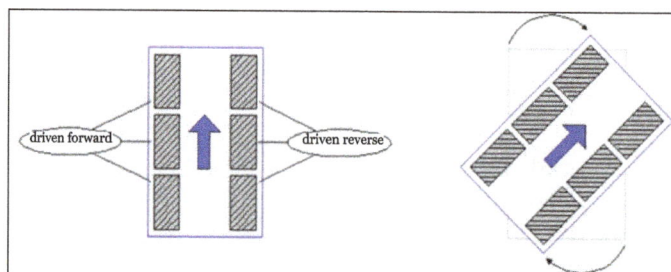

Skid Steer Drive.

Its properties are similar to the differential drive system, except that it provides greater traction and has quite inaccurate position tracking.

## Articulated Drive

Here, the body/chassis of the robot is deformed to produce rotatory motion, whereas the translatory motion is provided by the wheels. Two motors are required, one for translatory motion (wheels) and another to change pivot angle (for the linear actuator).

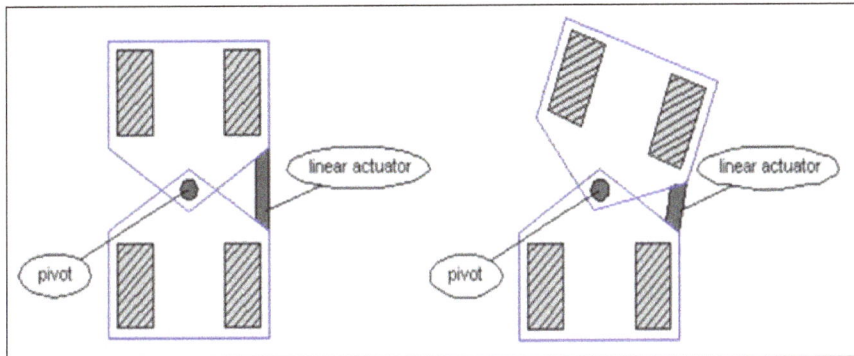

Articulated Drive.

Rest other features are similar to the car type drive.

## Synchronous Drive

Here, the robot can move in any direction without changing its alignment. Two sets of motors are required, one to drive the wheels, other to change their direction. This is clearly shown in the diagram above.

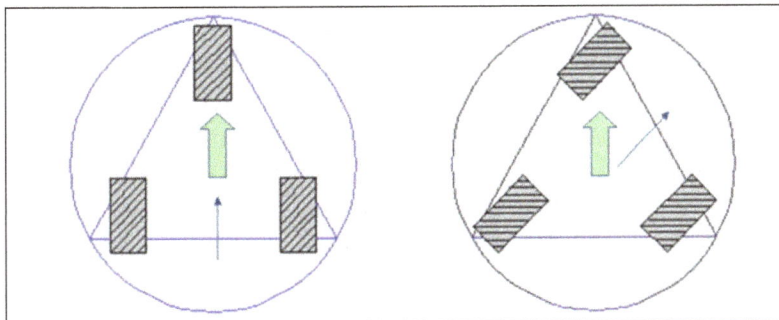

Synchronous Drive.

It has synchronous operation and greater accuracy. It is a bit complex to build, but simpler to use!

## Pivot Drive

This is a unique type of drive system. There is a four wheeled chassis which gives translatory motion and a rotating platform which gives rotational motion. Thus, it achieves accurate straight line motion. While turning, the raised platform is lowered such that it lifts the chassis, rotates by desired angle, and then is raised again to keep the chassis back on the ground. This can be achieved using one/two motors, depending upon the complexity and requirement.

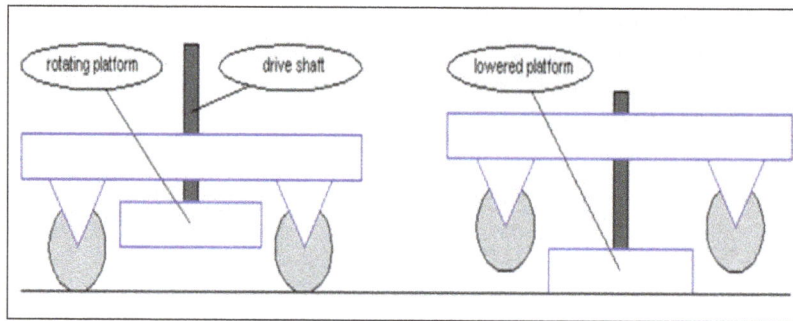

Pivot Drive.

## Dual Differential Drive

This is similar to differential drive, but uses special gear assemblies, which increase the accuracy of straight line motion and on-spot turning.

## Tracked Slip/Skid Locomotion

Wheeled locomotion offers some disadvantages, especially in case of omnidirectional vehicles using spherical or Swedish wheels, in rough, loose terrain, due to the increasing rolling friction which causes power inefficiencies as shown in figure; furthermore vehicles using wheels are just able to cross gaps that are smaller as the diameter of the vehicles wheels. In tracked slip/skid locomotion vehicles using tracks like a tank, one example of a robot using this concept is the Nanokhod robot which probably will go to mars. A tracked vehicle is steered by moving the tracks with different speed in the same direction or in opposite direction.

The use of tracks offers a much larger area of ground contact, so the vehicles traction on loose surface is much better than the traction of wheels, furthermore the vehicle is able to drive through rougher terrain than wheeled vehicles are (it is for example able to cross larger gaps). Due to the large contact patches, tracked vehicles usually change their direction by skidding, where a large part of the vehicle slides against the ground, so the vehicle needs a lot of space to change the orientation of the chassis. The skidding movement has some other disadvantages which are coupled with the steering method itself and the surface. When the surface is hard (for example a tarred road) the vehicle is not able to slide against it, this increases the friction during steering and with this the power consumption of the vehicle. Furthermore the exact change of the robot's chassis in position and orientation is hard to predict due to the sliding movement and changing ground friction.

Nanokhod, developed by Hoerner and Sulger GMBH and the Max Planck Institute.

Shrimp (EPFL).

## Walking Wheels

Legged robots are able to climb stairs and travel through rough terrain, but they offer some inefficiencies on flat surface and controlling the robots is difficult. Wheeled robots are very energy efficient on hard surface, even at high speed, but most of them are surely not able to climb stairs. One idea is a hybrid solution which combines the advantages of legged and wheeled locomotion. Figure shows shrimp, which was developed by the EPFL, a robot which uses 'walking wheels' to locomote. Shrimp has six motorized wheels and is capable to climb barriers that are two times larger than its wheel diameter. Shrimp has a steering wheel in the front and rear and two wheels arranged in a bogie at each side. Steering is realized by synchronizing the steering of the front and rear wheel and speed difference of the bogie wheels. This steering method allows high precision manoeuvres with a minimal skid movement of the four bogie wheels. One of the most interesting features of shrimp is that it is able to overcome obstacles passively, that means that the robot has no sensors to detect an obstacle, the robot's mechanical structure is able to adapt the profile of the terrain. Some interesting videos which are showing shrimp's abilities can be taken from.

# BIO-INSPIRED ROBOTIC LOCOMOTION

## Bio-inspired Robotics

Two u-CAT robots that are being developed at the Tallinn University
of Technology to reduce the cost of underwater archaeological operations.

Bio-inspired robotic locomotion is a fairly new subcategory of bio-inspired design. It is about learning concepts from nature and applying them to the design of real-world engineered systems. More specifically, this field is about making robots that are inspired by biological systems. Bio-mimicry and bio-inspired design are sometimes confused. Biomimicry is copying the nature while bio-inspired design is learning from nature and making a mechanism that is simpler and more effective than the system observed in nature. Biomimicry has led to the development of a different branch of robotics called soft robotics. The biological systems have been optimized for specific tasks according to their habitat. However, they are multifunctional and are not designed for only one specific functionality. Bio-inspired robotics is about studying biological systems, and look for the mechanisms that may solve a problem in the engineering field. The designer should then try to simplify and enhance that mechanism for the specific task of interest. Bio-inspired ro-boticists are usually interested in biosensors (e.g. eye), bioactuators (e.g. muscle), or biomaterials (e.g. spider silk). Most of the robots have some type of locomotion system. Thus, in this article different modes of animal locomotion and few examples of the corresponding bio-inspired robots are introduced.

Stickybot: a gecko-inspired robot.

## Biolocomotion

Biolocomotion or animal locomotion is usually categorized as below:

## Locomotion on a Surface

Locomotion on a surface may include terrestrial locomotion and arboreal locomotion. We will specifically discuss about terrestrial locomotionin detail in the next section.

Big eared townsend bat (Corynorhinus townsendii).

## Locomotion in a Fluid

Locomotion in a blood stream swimming and flying. There are many swimming and flying robots designed and built by roboticists.

## Behavioral Classification (Terrestrial Locomotion)

There are many animal and insects moving on land with or without legs. We will discuss legged and limbless locomotion in this section as well as climbing and jumping. Anchoring the feet is fundamental to locomotion on land. The ability to increase traction is important for slip-free motion on surfaces such as smooth rock faces and ice, and is especially critical for moving uphill. Numerous biological mechanisms exist for providing purchase: claws rely upon friction-based mechanisms; gecko feet upon van der walls forces; and some insect feet upon fluid-mediated adhesive forces.

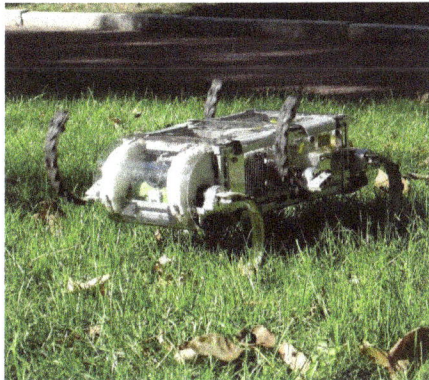

Rhex: a Reliable Hexapedal Robot.

## Legged Locomotion

Legged robots may have one, two, four, six, or many legs depending on the application. One of the main advantages of using legs instead of wheels is moving on uneven environment more effectively. Bipedal, quadrupedal, and hexapedal locomotion are among the most favorite types of legged locomotion in the field of bio-inspired robotics. Rhex, a Reliable Hexapedal robot and Cheetah are the two fastest running robots so far. iSprawl is another hexapedal robot inspired by cockroach locomotion that has been developed at Stanford University. This robot can run up to 15 body length per second and can achieve speeds of up to 2.3 m/s. The original version of this robot was pneumatically driven while the new generation uses a single electric motor for locomotion.

## Limbless Locomotion

Terrain involving topography over a range of length scales can be challenging for most organisms and biomimetic robots. Such terrain are easily passed over by limbless organisms such as snakes. Several animals and insects including worms, snails, caterpillars, and snakes are capable of limbless locomotion. A review of snake-like robots is presented by Hirose et al. These robots can be categorized as robots with passive or active wheels, robots with active treads, and undulating robots using vertical waves or linear expansions. Most snake-like robots use wheels, which are high in friction when moving side to side but low in friction when rolling forward (and can be prevented from rolling backward). The majority of snake-like robots use either lateral undulation or rectilinear locomotion and have difficulty climbing vertically. Choset has recently developed a modular robot that can mimic several snake gaits, but it cannot perform concertina motion. Researchers at Georgia Tech have recently developed two snake-like robots called Scalybot. The focus of these robots is on the role of snake ventral scales on adjusting the frictional properties in different directions. These robots can actively control their scales to modify their frictional properties and move on a variety of surfaces efficiently. Researchers at CMU have developed both scaled and conventional actuated snake-like robots.

## Climbing

Climbing is an especially difficult task because mistakes made by the climber may cause the climber to lose its grip and fall. Most robots have been built around a single functionality observed in their biological counterparts. Geckobots typically use van der waals forces that work only on smooth surfaces. Stickybots, and use directional dry adhesives that works best on smooth surfaces. Spiny-bot and the RiSE robot are among the insect-like robots that use spines instead. Legged climbing robots have several limitations. They cannot handle large obstacles since they are not flexible and they require a wide space for moving. They usually cannot climb both smooth and rough surfaces or handle vertical to horizontal transitions as well.

## Jumping

One of the tasks commonly performed by a variety of living organisms is jumping. Bharal, hares, kangaroo, grasshopper, flea, and locust are among the best jumping animals. A miniature 7g jumping robot inspired by locust has been developed at EPFL that can jump up to 138 cm. The jump event is induced by releasing the tension of a spring. The highest jumping miniature robot is inspired by the locust, weighs 23 grams with its highest jump to 365 cm is "TAUB" (Tel-Aviv University and Braude College of engineering). It uses torsion springs as energy storage and includes a wire and latch mechanism to compress and release the springs. ETH Zurich has reported a soft jumping robot based on the combustion of methane and laughing gas. The thermal gas expansion inside the soft combustion chamber drastically increases the chamber volume. This causes the 2 kg robot to jump up to 20 cm. The soft robot inspired by a roly-poly toy then reorientates itself into an upright position after landing.

## Behavioral Classification (Aquatic Locomotion)

## Swimming (Piscine)

It is calculated that when swimming some fish can achieve a propulsive efficiency greater than 90%. Furthermore, they can accelerate and maneuver far better than any man-made boat or submarine, and produce less noise and water disturbance. Therefore, many researchers studying underwater robots would like to copy this type of locomotion. Notable examples are the Essex University Computer Science Robotic Fish G9, and the Robot Tuna built by the Institute of Field Robotics, to analyze and mathematically model thunniform motion. The Aqua Penguin, designed and built by Festo of Germany, copies the streamlined shape and propulsion by front "flippers" of penguins. Festo have also built the Aqua Ray and Aqua Jelly, which emulate the locomotion of manta ray, and jellyfish, respectively.

Robotic Fish: iSplash-II.

In 2014, *iSplash*-II was developed by PhD student Richard James Clapham and Prof. Huosheng Hu at Essex University. It was the first robotic fish capable of outperforming real carangiform fish in terms of average maximum velocity (measured in body lengths/ second) and endurance, the duration that top speed is maintained. This build attained swimming speeds of 11.6BL/s (i.e. 3.7 m/s). The first build, *iSplash*-I (2014) was the first robotic platform to apply a full-body length carangiform swimming motion which was found to increase swimming speed by 27% over the traditional approach of a posterior confined waveform.

## Morphological Classification

## Modular

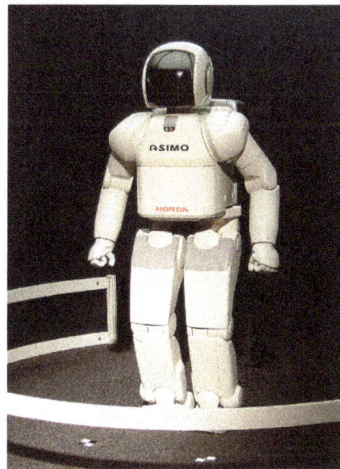
Honda Asimo: A Humanoid robot.

The modular robots are typically capable of performing several tasks and are specifically useful for search and rescue or exploratory missions. Some of the featured robots in this category include a salamander inspired robot developed at EPFL that can walk and swim, a snake inspired robot developed at Carnegie-Mellon University that has four different modes of terrestrial locomotion, and a cockroach inspired robot can run and climb on a variety of complex terrain.

## Humanoid

Humanoid robots are robots that look human-like or are inspired by the human form. There are many different types of humanoid robots for applications such as personal assistance, reception, work at industries, or companionship. These type of robots are used for research purposes as well and were originally developed to build better orthosis and prosthesis for human beings. Petman is one of the first and most advanced humanoid robots developed at Boston Dynamics. Some of the humanoid robots such as Honda Asimo are over actuated. On the other hand, there are some humanoid robots like the robot developed at Cornell University that do not have any actuators and walk passively descending a shallow slope.

## Swarming

The collective behavior of animals has been of interest to researchers for several years. Ants can make structures like rafts to survive on the rivers. Fish can sense their environment more effectively in large groups. Swarm robotics is a fairly new field and the goal is to make robots that can work together and transfer the data, make structures as a group, etc.

## Soft

Soft robots are robots composed entirely of soft materials and moved through pneumatic pressure, similar to an octopus or starfish. Such robots are flexible enough to move in very limited spaces (such as in the human body). The first multigait soft robots was developed in 2011 and the first fully integrated, independent soft robot (with soft batteries and control systems) was developed in 2015.

# OTHER METHODS OF LOCOMOTION

## Flying

Two robot snakes. Left one has 64 motors (with 2 degrees of freedom per segment), the right one 10.

A modern passenger airliner is essentially a flying robot, with two humans to manage it. The autopilot can control the plane for each stage of the journey, including takeoff, normal flight, and even landing. Other flying robots are uninhabited and are known as unmanned aerial vehicles(UAVs). They can be smaller and lighter without a human pilot on board, and fly into dangerous territory for military surveillance missions. Some can even fire on targets under command. UAVs are also being developed which can fire on targets automatically, without the need for a command from a human. Other flying robots include cruise missiles, the Entomopter, and the Epson micro helicopter robot. Robots such as the Air Penguin, Air Ray, and Air Jelly have lighter-than-air bodies, propelled by paddles, and guided by sonar.

## Snaking

Several snake robots have been successfully developed. Mimicking the way real snakes move, these robots can navigate very confined spaces, meaning they may one day be used to search for people trapped in collapsed buildings. The Japanese ACM-R5 snake robot can even navigate both on land and in water.

## Skating

A small number of skating robots have been developed, one of which is a multi-mode walking and skating device. It has four legs, with unpowered wheels, which can either step or roll. Another robot, Plen, can use a miniature skateboard or roller-skates, and skate across a desktop.

Capuchin, a climbing robot.

## Climbing

Several different approaches have been used to develop robots that have the ability to climb vertical surfaces. One approach mimics the movements of a human climber on a wall with protrusions; adjusting the center of mass and moving each limb in turn to gain leverage. An example of this is Capuchin, built by Dr. Ruixiang Zhang at Stanford University, California. Another approach uses the specialized toe pad method of wall-climbing geckoes, which can run on smooth surfaces such as vertical glass. Examples of this approach include Wallbot and Stickybot. China's *Technology Daily* reported on November 15, 2008, that Dr. Li Hiu Yeung and his research group of New Concept Aircraft (Zhuhai) Co., Ltd. had successfully developed a bionic gecko robot named "Speedy Freelander". According to Dr. Li, the gecko robot could rapidly climb up and down a variety of building walls, navigate through ground and wall fissures, and walk upside-down on the ceiling. It was also able to adapt to the surfaces of smooth glass, rough, sticky or dusty walls as well as various types of metallic materials. It could also identify and circumvent obstacles automatically. Its flexibility and speed were comparable to a natural gecko. A third approach is to mimic the motion of a snake climbing a pole.

## Swimming (Piscine)

It is calculated that when swimming some fish can achieve a propulsive efficiency greater than 90%.

Furthermore, they can accelerate and maneuver far better than any man-made boat or submarine, and produce less noise and water disturbance. Therefore, many researchers studying underwater robots would like to copy this type of locomotion. Notable examples are the Essex University Computer Science Robotic Fish G9, and the Robot Tuna built by the Institute of Field Robotics, to analyze and mathematically model thunniform motion. The Aqua Penguin, designed and built by Festo of Germany, copies the streamlined shape and propulsion by front "flippers" of penguins. Festo have also built the Aqua Ray and Aqua Jelly, which emulate the locomotion of manta ray, and jellyfish, respectively.

Robotic Fish: *iSplash*-II.

In 2014 *iSplash*-II was developed by PhD student Richard James Clapham and Prof. Huosheng Hu at Essex University. It was the first robotic fish capable of outperforming real carangiform fish in terms of average maximum velocity (measured in body lengths/ second) and endurance, the duration that top speed is maintained. This build attained swimming speeds of 11.6BL/s (i.e. 3.7 m/s). The first build, *iSplash*-I (2014) was the first robotic platform to apply a full-body length carangiform swimming motion which was found to increase swimming speed by 27% over the traditional approach of a posterior confined waveform.

## Sailing

The autonomous sailboat robot *Vaimos*.

Sailboat robots have also been developed in order to make measurements at the surface of the ocean. A typical sailboat robot is *Vaimos* built by IFREMER and ENSTA-Bretagne. Since the propulsion of sailboat robots uses the wind, the energy of the batteries is only used for the computer, for the communication and for the actuators (to tune the rudder and the sail). If the robot is equipped with solar panels, the robot could theoretically navigate forever. The two main competitions of sailboat robots are WRSC, which takes place every year in Europe, and Sailbot.

# ROBOTICS CONTROL SYSTEM

To perform as per the program instructions, the joint movements an industrial robot must accurately be controlled. Micro-processor-based controllers are used to control the robots. Different types of control that are being used in robotics are given as follows.

a. Limited Sequence Control

It is an elementary control type. It is used for simple motion cycles, such as pick-and-place operations. It is implemented by fixing limits or mechanical stops for each joint and sequencing the movement of joints to accomplish operation. Feedback loops may be used to inform the controller that the action has been performed, so that the program can move to the next step. Precision of such control system is less. It is generally used in pneumatically driven robots.

b. Playback with Point-to-Point Control

Playback control uses a controller with memory to record motion sequences in a work cycle, as well as associated locations and other parameters, and then plays back the work cycle during program execution. Point-to-point control means individual robot positions are recorded in the memory. These positions include both mechanical stops for each joint, and the set of values that represent locations in the range of each joint. Feedback control is used to confirm that the individual joints achieve the specified locations in the program.

c. Playback with Continuous Path Control

Continuous path control refers to a control system capable of continuous simultaneous control of two or more axes. The following advantages are noted with this type of playback control: greater storage capacity—the number of locations that can be stored is greater than in point-to-point; and interpolation calculations may be used, especially linear and circular interpolations.

d. Intelligent Control

An intelligent robot exhibits behavior that makes it seems to be intelligent. For example, it may have capacity to interact with its ambient surroundings; decision-making capability; ability to communicate with humans; ability to carry out computational analysis during the work cycle; and responsiveness to advanced sensor inputs. They may also possess the playback facilities. However it requires a high level of computer control, and an advanced programming language to input the decision-making logic and other 'intelligence' into the memory.

## End Effectors

An end effector is usually attached to the robot's wrist, and it allows the robot to accomplish a specific task. This means that end effectors are generally custom-engineered and fabricated for each different operation. There are two general categories of end effectors viz. grippers and tools.

Grippers grasp and manipulate the objects during the work cycle. Typically objects that grasped are the work parts which need to be loaded or unloaded from one station to another. Grippers may be custom-designed to suit the physical specifications of work parts. Various end-effectors, grippers are summarized in table.

Table End Effectors: Grippers.

| Type | Description |
|---|---|
| Mechanical | Two or more fingers which are actuated by robot controller to open and close on workpart. |
| Vacuum gripper | Suction cups are used to hold flat objects. |
| Magnetized devices | Based on the principle of magnetism. These are used to holding ferrous workparts. |
| Adhesive devices | By deploying adhesive substances, these are used to hold flexible materials, such as fabric. |
| Simple mechanical devices | Hooks and scoops. |
| Dual grippers | It is mechanical gripper with two gripping devices in one end-effecter. It is used for machine loading and unloading. It reduces cycle time per part by gripping two workparts at the same time. |
| Interchangeable fingers | Mechanical gripper with an arrangement to have modular fingers to accommodate different sizes workpart. |
| Sensory feedback fingers | Mechanical gripper with sensory feedback capabilities in the fingers to aid locting the workpart; and to determine correct grip force to apply (for fragile workparts). |
| Multiple fingered | Mechanical gripper as per the general anatomy of human hand. |
| Standard grippers | Mechanical gripper that are commercially available, thus reducing the need to custom-design a gripper for separate robot applications. |

The robot end effecter may also use tools. Tools are used to perform processing operations on the workpart. Typically the robot uses the tool relative to a stationary or slowly-moving object. For example, spot welding, arc welding, and spray painting robots use a tool for processing the respective operation. Tools also can be mounted at robotic manipulator spindle to carry out machining work such as drilling, routing, grinding, etc.

## Sensors in Robotics

There are generally two categories of sensors used in robotics. These are sensors for internal purposes and for external purposes. Internal sensors are used to monitor and control the various joints of the robot. They form a feedback control loop with the robot controller. Examples of internal sensors include potentiometers and optical encoders, while tachometers of various types are deployed to control the speed of the robot arm.

External sensors are external to the robot itself, and are used when we wish to control the operations of the robot. External sensors are simple devices, such as limit switches that determine whether a part has been positioned properly, or whether a part is ready to be picked up from an unloading bay.

Various sensors used in robotics are outlined in table.

Table: Sensor Technologies for Robotics.

| Sensor type | Description |
|---|---|
| Tactile sensors | Used to determine whether contact is made between sensor and another object.<br><br>Touch sensors: indicates the contact.<br><br>Force sensors: indicates the magnitude of force with the object. |
| Proximity sensors | Used to determine how close an object is to the sensor. Also called a range sensor. |
| Optical sensors | Photocells and other photometric devices that are used to detect the presence or absence of objects. Often used in conjunction with proximity sensors. |
| Machine vision | Used in robotics for inspection, parts identification, guidance, etc. |
| Others | Measurement of temperature, fluid pressure, fluid flow, electrical voltage, current, and other physical properties. |

## Industrial Robot Applications

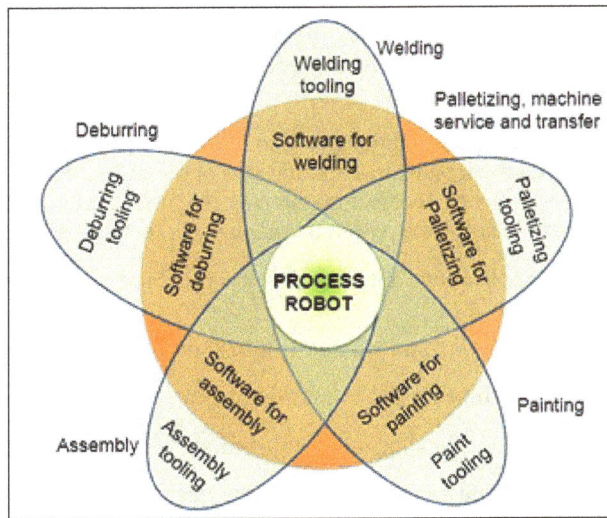

Applications of robots in industry and manufacturing.

Figure shows a diagram which depicts an overview of applications of robots in manufacturing. The general characteristics of industrial work situations that tend to promote the substitution of robots for human labor are outlined in Table.

Table: Characteristics of situations where robots may substitute for humans.

| Situation | Description |
|---|---|
| Hazardous work environment for humans | In situations where the work environment is unsafe, unhealthy, uncomfortable, or otherwise unpleasant for humans, robot application may be considered. |
| Repetitive work cycle | If the sequence of elements in the work cycle is the same, and the elements consist of relatively simple motions, robots usually perform the work with greater consistency and repeatability than humans. |
| Difficult handling for humans | If the task requires the use of heavy or difficult-to-handle parts or tools for humans, robots may be able to perform the operation more efficiently. |
| Multi-shift operation | A robot can replace two or three workers at a time in second or third shifts, thus they can provide a faster financial payback. |

| Infrequent changeovers | Robots' use is justified for long production runs where there are infrequent changeovers, as opposed to batch or job shop production where changeovers are more frequent. |
|---|---|
| Part position and orientation are established in the work cell | Robots generally don't have vision capabilities, which means parts must be precisely placed and oriented for successful robotic opetations. |

## Material Handling Applications

Robots are mainly used in three types of applications: material handling; processing operations; and assembly and inspection. In material handling, robots move parts between various locations by means of a gripper type end effector. Material handling activity can be sub divided into material transfer and machine loading and/or unloading. These are described in Table.

Table: Material handling applications.

| Application | Description |
|---|---|
| Material transfer | • Main purpose is to pick up parts at one location and place them at a new location. Part re-orientation may be accomplished during the transfer. The most basic application is pick-and place procedure, by a low-technology robot (often pneumatic), using only up to 4 joints.<br><br>• More complex is palletizing, where robots retrieve objects from one location, and deposit them on pallet in a specific area of the pallet, thus the deposit location is slightly different for each object transferred. The robot must be able to compute the correct deposit location via powered lead-through method, or by dimensional analysis.<br><br>• Other applications of material transfer include de-palletizing, stacking, and insertion operations. |
| Machine loading and/or unloading | • Primary aim is to transfer parts into or out of a production machine.<br><br>• There are three classes to consider:<br><br>    ○ Machine loading - where the robot loads the machine<br><br>    ○ Machine unloading – where the robot unloads the machine<br><br>    ○ Machine loading and unloading - where the robot performs both actions<br><br>• Used in die casting, plastic molding, metal machining operations, forging, press-working, and heat treating operations. |

## Processing Operations

In processing operations, the robot performs some processing activities such as grinding, milling, etc. on the workpart. The end effector is equipped with the specialized tool required for the respective process. The tool is moved relative to the surface of the workpart. Table outlines the examples of various processing operations that deploy robots.

Table: Robotic process operations.

| Process | Description |
|---|---|
| Spot welding | Metal joining process in which two sheet metal parts are fused together at localized points of contact by the deployment of two electrodes that squeeze the metal together and apply an electric current. The electrodes constitute the spot welding gun, which is the end effector tool of the wilding robot. |

| Arc welding | Metal joining process that utilizes a continuous rather than contact welding point process, in the same way as above. again the end effector is the electrodes used to achieve the welding arc. The robot must use continuous path control, and a jointed arm robot consisting of six joints is frequently used. |
|---|---|
| Spray coating | Spray coating directs a spray gun at the object to be coated. Paint or some other fluid flows through the nozzle of the spray gun, which is the end effector, and is dispersed and applied over the surface of the object. Again the robot must use continuous path control, and is typically programmed using manual lead-through. Jointed arm robots seem to be most common anatomy for this application. |
| Other applications | Other applications include: drilling routing and other machining processes; grinding wire brushing, and similar operations; waterjet cutting; and laser cutting. |

## Robot Programming

A robot program is a path in the space that to be followed by the manipulator, combined with peripheral actions that support the work cycle. To program a robot, specific commands are entered into the robot's controller memory, and these actions may be performed in a number of ways. Limited sequence robot programming is carried out when limit switches and mechanical stops are set to control the end-points of its motions. A sequencing device controls the occurrence of motions, which in turn controls the movement of the joints that completes the motion cycle.

## Lead-through Programming

For industrial robots with digital computers as controllers, three programming methods can be distinguished. These are lead-through programming; computer-like robot programming languages; and off-line programming. Lead-through methodologies, and associated programming methods, are outlined in Table.

Table: Lead-through programming methods for industrial robots.

| Method | Description |
|---|---|
| Lead-through programming | • Task is taught to the robot by manually moving the manipulator through the required motion cycle, and simultaneously entering the program into the controller memory for playback.<br><br>• Two methods are used for teaching: powered lead-through; and manual lead-through. |
| Motion programming | • To overcome the difficulties of co-coordinating individual joints associated with lead-through programming, two mechanical methods can be used: the world-co-ordinate system-whereby the origin and axes are defined relative to the robot base; and the tool-co-ordinate system-whereby the alignment of the axis system is defined relative to the orientation of the wrist faceplate.<br><br>• These methods are typically used with cartesian co-ordinate robots, and not for robots with rotational joints.<br><br>• Robotic types with rotational joints rely on interpolation processes to gain straight line motion. |

| | • Two types of interpolation processes are used : straight line interpolation-where the control computer calculates the necessary points in space that the manipulator must move through to connect two points; in space that the manipulator must move through to connect two points; and joint interpolation-where joints are moved simultaneously at their own constant speed such that all joints start/stop at the same time. |
|---|---|

## Computer-like Programming

These are computer-like languages which use on-line/off-line methods of programming. The advantages of textual programming over its lead-through counterpart include:

- The use of enhanced sensor capabilities, including the use of analogue and digital inputs.

- Improved output capabilities for controlling external equipment.

- Extended program logic, beyond lead-through capabilities.

- Advanced computation and data processing capabilities.

- Communications with other computer systems.

# MOTION PLANNING

Motion planning (also known as the navigation problem or the piano mover's problem) is a term used in robotics is to find a sequence of valid configurations that moves the robot from the source to destination.

For example, consider navigating a mobile robot inside a building to a distant waypoint. It should execute this task while avoiding walls and not falling down stairs. A motion planning algorithm would take a description of these tasks as input, and produce the speed and turning commands sent to the robot's wheels. Motion planning algorithms might address robots with a larger number of joints (e.g., industrial manipulators), more complex tasks (e.g. manipulation of objects), different constraints (e.g., a car that can only drive forward), and uncertainty (e.g. imperfect models of the environment or robot).

Motion planning has several robotics applications, such as autonomy, automation, and robot design in CAD software, as well as applications in other fields, such as animating digital characters, video game, artificial intelligence, architectural design, robotic surgery, and the study of biological molecules.

## Concepts

A basic motion planning problem is to produce a continuous motion that connects a start configuration S and a goal configuration G, while avoiding collision with known obstacles. The robot and obstacle geometry is described in a 2D or 3D *workspace*, while the motion is represented as a path in (possibly higher-dimensional) configuration space.

Example of a workspace.

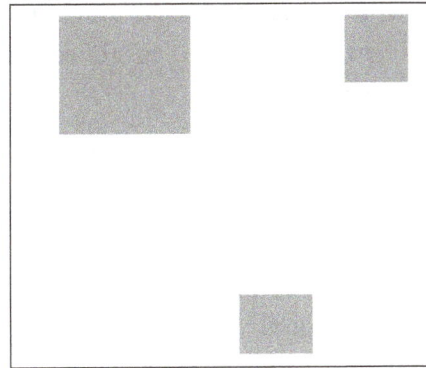

Configuration space of a point-sized robot.
White = *Cfree*, gray = *Cobs*.

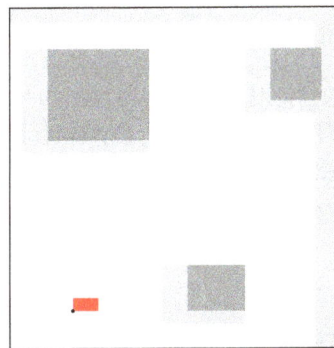

Configuration space for a rectangular translating robot (pictured red). White = *Cfree*, gray = *Cobs*, where dark gray = the objects, light gray = configurations where the robot would touch an object or leave the workspace.

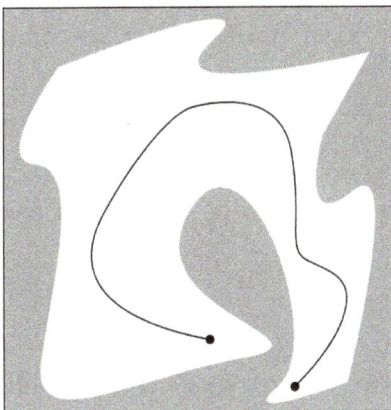

Example of a valid path.

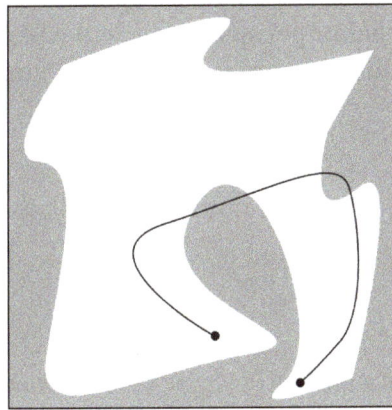

Example of an invalid path.

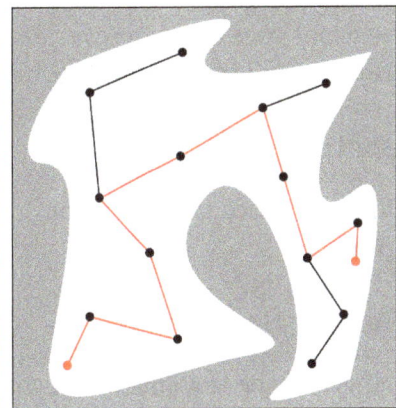

Example of a road map.

## Configuration Space

A configuration describes the pose of the robot, and the configuration space C is the set of all possible configurations. For example:

- If the robot is a single point (zero-sized) translating in a 2-dimensional plane (the workspace), C is a plane, and a configuration can be represented using two parameters (x, y).

- If the robot is a 2D shape that can translate and rotate, the workspace is still 2-dimensional.

However, C is the special Euclidean group SE(2) = R2 × SO(2) (where SO(2) is the special orthogonal group of 2D rotations), and a configuration can be represented using 3 parameters (x, y, θ).

- If the robot is a solid 3D shape that can translate and rotate, the workspace is 3-dimensional, but C is the special Euclidean group **SE(3) = R3 × SO(3)**, and a configuration requires 6 parameters: (x, y, z) for translation, and Euler angles (α, β, γ).

- If the robot is a fixed-base manipulator with N revolute joints (and no closed-loops), C is N-dimensional.

## Free Space

The set of configurations that avoids collision with obstacles is called the free space Cfree. The complement of Cfree in C is called the obstacle or forbidden region.

Often, it is prohibitively difficult to explicitly compute the shape of Cfree. However, testing whether a given configuration is in Cfree is efficient. First, forward kinematics determine the position of the robot's geometry, and collision detectiontests if the robot's geometry collides with the environment's geometry.

## Target Space

Target space is a linear subspace of free space which denotes where we want the robot to move to. In global motion planning, target space is observable by the robot's sensors. However, in local motion planning, the robot cannot observe the target space in some states. To solve this problem, the robot goes through several virtual target spaces, each of which is located within the observable area (around the robot). A virtual target space is called a sub-goal.

## Obstacle Space

Obstacle space is a space that the robot can not move to. Obstacle space **is not** opposite of free space.

## Danger Space

Danger space is a space that robot can but it is not desired move to. Danger space is not an obstacle nor free space. For example, mud pits in a wooded area and greasy floor in a factory can be considered as a danger space. If the robot cannot find a trajectory that completely belong to free space, it must pass through the danger space. In some cases, in the motion planning with time/energy constraint robot may prefer to take a short trajectory in danger space instead of long path in free space.

## Algorithms

Low-dimensional problems can be solved with grid-based algorithms that overlay a grid on top of configuration space, or geometric algorithms that compute the shape and connectivity of Cfree.

Exact motion planning for high-dimensional systems under complex constraints is computationally intractable. Potential-field algorithms are efficient, but fall prey to local minima (an exception is the harmonic potential fields). Sampling-based algorithms avoid the problem of local minima, and solve many problems quite quickly. They are unable to determine that no path exists, but they have a probability of failure that decreases to zero as more time is spent.

Sampling-based algorithms are currently considered state-of-the-art for motion planning in high-dimensional spaces, and have been applied to problems which have dozens or even hundreds of dimensions (robotic manipulators, biological molecules, animated digital characters, and legged robots).

## Grid-based Search

Grid-based approaches overlay a grid on configuration space, and assume each configuration is identified with a grid point. At each grid point, the robot is allowed to move to adjacent grid points as long as the line between them is completely contained within Cfree (this is tested with collision detection). This discretizes the set of actions, and search algorithms (like A*) are used to find a path from the start to the goal.

These approaches require setting a grid resolution. Search is faster with coarser grids, but the algorithm will fail to find paths through narrow portions of Cfree. Furthermore, the number of points on the grid grows exponentially in the configuration space dimension, which make them inappropriate for high-dimensional problems.

Traditional grid-based approaches produce paths whose heading changes are constrained to multiples of a given base angle, often resulting in suboptimal paths. Any-angle path planning approaches find shorter paths by propagating information along grid edges (to search fast) without constraining their paths to grid edges (to find short paths).

Grid-based approaches often need to search repeatedly, for example, when the knowledge of the robot about the configuration space changes or the configuration space itself changes during path following. Incremental heuristic search algorithms replan fast by using experience with the previous similar path-planning problems to speed up their search for the current one.

## Interval-based Search

These approaches are similar to grid-based search approaches except that they generate a paving covering entirely the configuration space instead of a grid. The paving is decomposed into two subpavings X−,X+ made with boxes such that X− ⊂ Cfree ⊂ X+. Characterizing Cfree amounts to solve a set inversion problem. Interval analysis could thus be used when Cfree cannot be described by linear inequalities in order to have a guaranteed enclosure.

The robot is thus allowed to move freely in X−, and cannot go outside X+. To both subpavings, a neighbor graph is built and paths can be found using algorithms such as Dijkstra or A*. When a path is feasible in X−, it is also feasible in Cfree. When no path exists in X+ from one initial configuration to the goal, we have the guarantee that no feasible path exists in Cfree. As for the grid-based approach, the interval approach is inappropriate for high-dimensional problems, due to the fact that the number of boxes to be generated grows exponentially with respect to the dimension of configuration space.

An illustration is provided by the three figures on the right where a hook with two degrees of freedom has to move from the left to the right, avoiding two horizontal small segments.

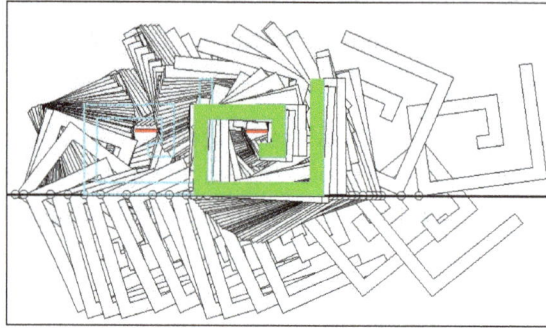

Motion from the initial configuration (blue) to the final configuration of the hook, avoiding the two obstacles (red segments). The left-bottom corner of the hook has to stay on the horizontal line, which makes the hook two degrees of freedom.

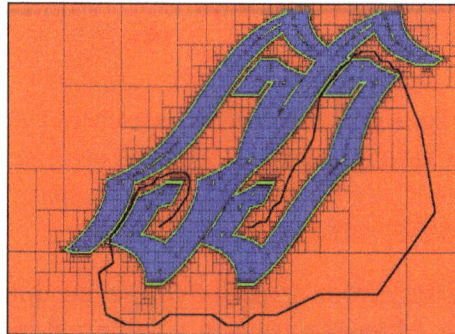

Decomposition with boxes covering the configuration space: The subpaving X− is the union all red boxes and the subpaving X+ is the union of red and green boxes. The path corresponds to the motion represented above.

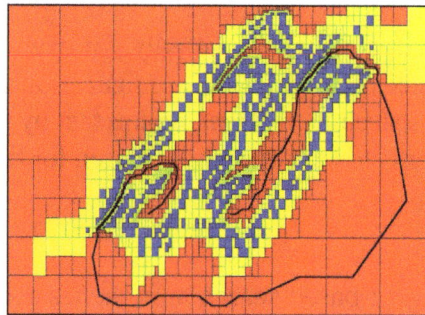

This figure corresponds to the same path as above but obtained with many fewer boxes.The algorithm avoids bisecting boxes in parts of the configuration space that do not influence the final result.

The decomposition with subpavings using interval analysis also makes it possible to characterize the topology of Cfreesuch as counting its number of connected components.

## Geometric Algorithms

Point robots among polygonal obstacles:

- Visibility graph.

- Cell decomposition.

Translating objects among obstacles:

- Minkowski sum.

Finding the way out of a building:

- Farthest ray trace.

Given a bundle of rays around the current position attributed with their length hitting a wall, the robot moves into the direction of the longest ray unless a door is identified. Such an algorithm was used for modeling emergency egress from buildings.

## Reward-based Algorithms

Reward-based algorithms assume that the robot in each state (position and internal state, including direction) can choose between different actions (motion). However, the result of each action is not definite. In other words, outcomes (displacement) are partly random and partly under the control of the robot. The robot gets positive reward when it reaches the target and gets negative reward if it collides with an obstacle. These algorithms try to find a path which maximizes cumulative future rewards. The Markov decision process (MDP) is a popular mathematical framework that is used in many reward-based algorithms. The advantage of MDPs over other reward-based algorithms is that they generate the optimal path. The disadvantage of MDPs is that they limit the robot to choose from a finite set of actions. Therefore, the path is not smooth (similar to grid-based approaches). Fuzzy Markov decision processes (FDMPs)is an extension of MDPs which generate smooth path with using an fuzzy inference system.

## Artificial Potential Fields

One approach is to treat the robot's configuration as a point in a potential field that combines attraction to the goal, and repulsion from obstacles. The resulting trajectory is output as the path. This approach has advantages in that the trajectory is produced with little computation. However, they can become trapped in local minima of the potential field and fail to find a path, or can find a non-optimal path. The artificial potential fields can be treated as continuum equations similar to electrostatic potential fields (treating the robot like a point charge), or motion through the field can be discretized using a set of linguistic rules.

## Sampling-based Algorithms

Sampling-based algorithms represent the configuration space with a roadmap of sampled configurations. A basic algorithm samples N configurations in C, and retains those in Cfree to use as *milestones*. A roadmap is then constructed that connects two milestones P and Q if the line segment PQ is completely in Cfree. Again, collision detection is used to test inclusion in Cfree. To find a path that connects S and G, they are added to the roadmap. If a path in the roadmap links S and G, the planner succeeds, and returns that path. If not, the reason is not definitive: either there is no path in Cfree, or the planner did not sample enough milestones.

These algorithms work well for high-dimensional configuration spaces, because unlike combinatorial algorithms, their running time is not (explicitly) exponentially dependent on the dimension of

C. They are also (generally) substantially easier to implement. They are probabilistically complete, meaning the probability that they will produce a solution approaches 1 as more time is spent. However, they cannot determine if no solution exists.

Given basic *visibility* conditions on Cfree, it has been proven that as the number of configurations N grows higher, the probability that the above algorithm finds a solution approaches 1 exponentially. Visibility is not explicitly dependent on the dimension of C; it is possible to have a high-dimensional space with "good" visibility or a low-dimensional space with "poor" visibility. The experimental success of sample-based methods suggests that most commonly seen spaces have good visibility.

There are many variants of this basic scheme:

- It is typically much faster to only test segments between nearby pairs of milestones, rather than all pairs.

- Nonuniform sampling distributions attempt to place more milestones in areas that improve the connectivity of the roadmap.

- Quasirandom samples typically produce a better covering of configuration space than pseudorandom ones, though some recent work argues that the effect of the source of randomness is minimal compared to the effect of the sampling distribution.

- It is possible to substantially reduce the number of milestones needed to solve a given problem by allowing curved eye sights (for example by crawling on the obstacles that block the way between two milestones).

- If only one or a few planning queries are needed, it is not always necessary to construct a roadmap of the entire space. Tree-growing variants are typically faster for this case (single-query planning). Roadmaps are still useful if many queries are to be made on the same space (multi-query planning).

## List of Notable Algorithms

- A*

- D*

- Rapidly-exploring random tree

- Probabilistic roadmap

## Completeness and Performance

A motion planner is said to be complete if the planner in finite time either produces a solution or correctly reports that there is none. Most complete algorithms are geometry-based. The performance of a complete planner is assessed by its computational complexity.

*Resolution completeness* is the property that the planner is guaranteed to find a path if the resolution of an underlying grid is fine enough. Most resolution complete planners are grid-based or interval-based. The computational complexity of resolution complete planners is dependent on the number of points in the underlying grid, which is $O(1/hd)$, where h is the resolution (the length of one side of a grid cell) and d is the configuration space dimension.

*Probabilistic completeness* is the property that as more "work" is performed, the probability that the planner fails to find a path, if one exists, asymptotically approaches zero. Several sample-based methods are probabilistically complete. The performance of a probabilistically complete planner is measured by the rate of convergence.

*Incomplete* planners do not always produce a feasible path when one exists. Sometimes incomplete planners do work well in Practice.

## Problem Variants

Many algorithms have been developed to handle variants of this basic problem.

## Differential Constraints

Holonomic:

- Manipulator arms (with dynamics).

Nonholonomic:

- Cars.

- Unicycles.

- Planes.

- Acceleration bounded systems.

- Moving obstacles (time cannot go backward).

- Bevel-tip steerable needle.

- Differential Drive Robots.

## Optimality Constraints

### Hybrid Systems

Hybrid systems are those that mix discrete and continuous behavior. Examples of such systems are:

- Robotic manipulation.

- Mechanical assembly.

- Legged robot locomotion.

- Reconfigurable robots.

## Uncertainty

- Motion uncertainty.

- Missing information.

- Active sensing.

- Sensorless planning.

## Applications

- Robot navigation.

- Automation.

- The driverless car.

- Robotic surgery.

- Digital character animation.

- Protein folding.

- Safety and accessibility in computer-aided architectural design.

# VISUAL SERVOING

Visual servoing, also known as vision-based robot control and abbreviated VS, is a technique which uses feedback information extracted from a vision sensor (visual feedback) to control the motion of a robot. One of the earliest papers that talks about visual servoing was from the SRI International Labs in 1979.

## Visual Servoing Taxonomy

There are two fundamental configurations of the robot end-effector (hand) and the camera:

- Eye-in-hand, or end-point closed-loop control, where the camera is attached to the moving hand and observing the relative position of the target.

- Eye-to-hand, or end-point open-loop control, where the camera is fixed in the world and observing the target and the motion of the hand.

Visual Servoing control techniques are broadly classified into the following types:

- Image-based (IBVS).

- Position/pose-based (PBVS).

- Hybrid approach.

IBVS was proposed by Weiss and Sanderson. The control law is based on the error between current and desired features on the image plane, and does not involve any estimate of the pose of the target. The features may be the coordinates of visual features, lines or moments of regions. IBVS has difficulties with motions very large rotations, which has come to be called camera retreat.

PBVS is a model-based technique (with a single camera). This is because the pose of the object of interest is estimated with respect to the camera and then a command is issued to the robot controller, which in turn controls the robot. In this case the image features are extracted as well, but are additionally used to estimate 3D information (pose of the object in Cartesian space), hence it is servoing in 3D.

Hybrid approaches use some combination of the 2D and 3D servoing. There have been a few different approaches to hybrid servoing:

- 2-1/2-D Servoing.

- Motion partition based.

- Partitioned DOF based.

### Software

- Matlab toolbox for visual servoing.

- Java-based visual servoing simulator.

- ViSP (ViSP states for "Visual Servoing Platform") is a modular software that allows fast development of visual servoing applications.

# TELEROBOTICS

Justus security robot patrolling in Kraków.

Telerobotics is the area of robotics concerned with the control of semi-autonomous robots from a distance, chiefly using Wireless network (like Wi-Fi, Bluetooth, the Deep Space Network, and similar) or tethered connections. It is a combination of two major subfields, teleoperation and telepresence.

# Teleoperation

Teleoperation indicates operation of a machine at a distance. It is similar in meaning to the phrase "remote control" but is usually encountered in research, academic and technical environments. It is most commonly associated with robotics and mobile robots but can be applied to a whole range of circumstances in which a device or machine is operated by a person from a distance.

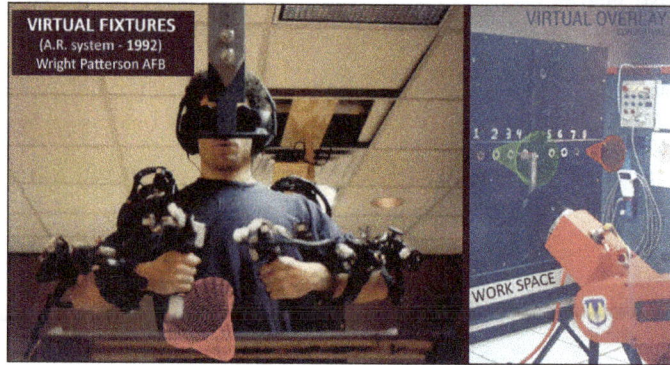

Early Telerobotics US Air Force - Virtual Fixtures system.

Teleoperation is the most standard term, used both in research and technical communities, for referring to operation at a distance. This is opposed to "telepresence", which refers to the subset of telerobotic systems configured with an immersive interface such that the operator feels present in the remote environment, projecting his or her presence through the remote robot. One of the first telepresence systems that enabled operators to feel present in a remote environment through all of the primary senses (sight, sound, and touch) was the Virtual Fixtures system developed at US Air Force Research Laboratories in the early 1990s. The system enabled operators to perform dexterous tasks (inserting pegs into holes) remotely such that the operator would feel as if he or she was inserting the pegs when in fact it was a robot remotely performing the task.

A telemanipulator (or teleoperator) is a device that is controlled remotely by a human operator. In simple cases the controlling operator's command actions correspond directly to actions in the device controlled, as for example in a radio controlled model aircraft or a tethered deep submergence vehicle. Where communications delays make direct control impractical (such as a remote planetary rover), or it is desired to reduce operator workload (as in a remotely controlled spy or attack aircraft), the device will not be controlled directly, instead being commanded to follow a specified path. At increasing levels of sophistication the device may operate somewhat independently in matters such as obstacle avoidance, also commonly employed in planetary rovers.

Devices designed to allow the operator to control a robot at a distance are sometimes called telecheric robotics.

Two major components of telerobotics and telepresence are the visual and control applications. A remote camera provides a visual representation of the view from the robot. Placing the robotic camera in a perspective that allows intuitive control is a recent technique that although based in Science Fiction (Robert A. Heinlein's Waldo 1942) has not been fruitful as the speed, resolution and bandwidth have only recently been adequate to the task of being able to control the robot camera in a meaningful way. Using a head mounted display, the control of the camera can be facilitated by tracking the head as shown in the figure below.

This only works if the user feels comfortable with the latency of the system, the lag in the response to movements, the visual representation. Any issues such as, inadequate resolution, latency of the video image, lag in the mechanical and computer processing of the movement and response, and optical distortion due to camera lens and head mounted display lenses, can cause the user 'simulator sickness' that is exacerbated by the lack of vestibular stimulation with visual representation of motion.

Mismatch between the users motions such as registration errors, lag in movement response due to overfiltering, inadequate resolution for small movements, and slow speed can contribute to these problems.

The same technology can control the robot, but then the eye–hand coordination issues become even more pervasive through the system, and user tension or frustration can make the system difficult to use.

The tendency to build robots has been to minimize the degrees of freedom because that reduces the control problems. Recent improvements in computers has shifted the emphasis to more degrees of freedom, allowing robotic devices that seem more intelligent and more human in their motions. This also allows more direct teleoperation as the user can control the robot with their own motions.

## Interfaces

A telerobotic interface can be as simple as a common MMK (monitor-mouse-keyboard) interface. While this is not immersive, it is inexpensive. Telerobotics driven by internet connections are often of this type. A valuable modification to MMK is a joystick, which provides a more intuitive navigation scheme for planar robot movement.

Dedicated telepresence setups utilize a head mounted display with either single or dual eye display, and an ergonomically matched interface with joystick and related button, slider, trigger controls.

Other interfaces merge fully immersive virtual reality interfaces and real-time video instead of computer-generated images. Another example would be to use an omnidirectional treadmill with an immersive display system so that the robot is driven by the person walking or running. Additional modifications may include merged data displays such as Infrared thermal imaging, real-time threat assessment, or device schematics.

## Applications

### Space

With the exception of the Apollo program, most space exploration has been conducted with terobotic space probes. Most space-based astronomy, for example, has been conducted with telerobotic telescopes. The Russian Lunokhod-1 mission, for example, put a remotely driven rover on the moon, which was driven in real time (with a 2.5-second lightspeed time delay) by human operators on the ground. Robotic planetary exploration programs use spacecraft that are programmed by humans at ground stations, essentially achieving a long-time-delay form of telerobotic operation. Recent noteworthy examples include the Mars exploration rovers (MER) and the Curiosity rover.

In the case of the MER mission, the spacecraft and the rover operated on stored programs, with the rover drivers on the ground programming each day's operation. The International Space Station (ISS) uses a two-armed telemanipulator called Dextre. More recently, a humanoid robot Robonaut has been added to the space station for telerobotic experiments.

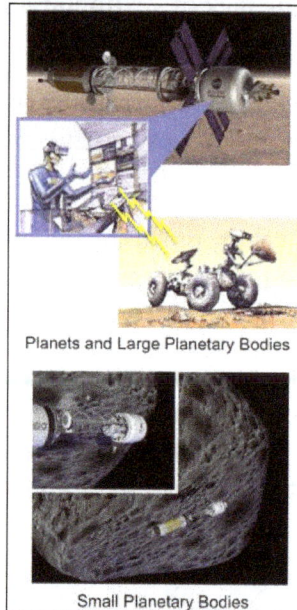

NASA HERRO (Human Exploration using Real-time Robotic Operations) telerobotic exploration concept.

NASA has proposed use of highly capable telerobotic systems for future planetary exploration using human exploration from orbit. In a concept for Mars Exploration proposed by Landis, a precursor mission to Mars could be done in which the human vehicle brings a crew to Mars, but remains in orbit rather than landing on the surface, while a highly capable remote robot is operated in real time on the surface. Such a system would go beyond the simple long time delay robotics and move to a regime of virtual telepresence on the planet. One study of this concept, the Human Exploration using Real-time Robotic Operations (HERRO) concept, suggested that such a mission could be used to explore a wide variety of planetary destinations.

## Telepresence and Videoconferencing

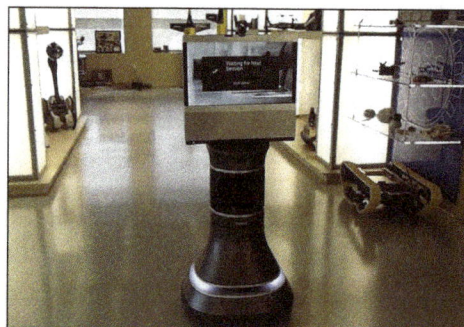

iRobot Ava 500, an autonomous roaming telepresence robot.

The prevalence of high quality video conferencing using mobile devices, tablets and portable computers has enabled a drastic growth in telepresence robots to help give a better sense of remote physical presence for communication and collaboration in the office, home, school, etc. when one cannot be there in person. The robot avatar can move or look around at the command of the remote person.

There have been two primary approaches that both utilize videoconferencing on a display 1) desktop telepresence robots - typically mount a phone or tablet on a motorized desktop stand to enable the remote person to look around a remote environment by panning and tilting the display or 2) drivable telepresence robots - typically contain a display (integrated or separate phone or tablet) mounted on a roaming base. Some examples of desktop telepresence robots include Kubi by Revolve Robotics, Galileo by Motrr, and Swivl. Some examples of roaming telepresence robots include Beam by Suitable Technologies, Double by Double Robotics, RP-Vita by iRobot and InTouch Health, Anybots, Vgo, TeleMe by Mantarobot, and Romo by Romotive. More modern roaming telepresence robots may include an ability to operate autonomously. The robots can map out the space and be able to avoid obstacles while driving themselves between rooms and their docking stations.

Traditional videoconferencing systems and telepresence rooms generally offer Pan / Tilt / Zoom cameras with far end control. The ability for the remote user to turn the device's head and look around naturally during a meeting is often seen as the strongest feature of a telepresence robot. For this reason, the developers have emerged in the new category of desktop telepresence robots that concentrate on this strongest feature to create a much lower cost robot. The desktop telepresence robots, also called head and neck Robots allow users to look around during a meeting and are small enough to be carried from location to location, eliminating the need for remote navigation.

Some telepresence robots are highly helpful for some long-term illness children, who were unable to attend school regularly. Latest innovative technologies can bring people together, and it allows them to stay connected to each other, which significantly help them to overcome loneliness.

## Marine Applications

Marine remotely operated vehicles (ROVs) are widely used to work in water too deep or too dangerous for divers. They repair offshore oil platforms and attach cables to sunken ships to hoist them. They are usually attached by a tether to a control center on a surface ship. The wreck of the *Titanic* was explored by an ROV, as well as by a crew-operated vessel.

## Telemedicine

Additionally, a lot of telerobotic research is being done in the field of medical devices, and minimally invasive surgical systems. With a robotic surgery system, a surgeon can work inside the body through tiny holes just big enough for the manipulator, with no need to open up the chest cavity to allow hands inside.

## Other Applications

Remote manipulators are used to handle radioactive materials.

Telerobotics has been used in installation art pieces; Telegarden is an example of a project where a robot was operated by users through the Web.

# ADAPTIVE COLLABORATIVE CONTROL

Adaptive collaborative control is the decision-making approach used in hybrid models consisting of finite-state machines with functional models as subcomponents to simulate behavior of systems formed through the partnerships of multiple agents for the execution of tasks and the development of work products. The term "collaborative control" originated from work developed in the late 90's and early 2000 by Fong, Thorpe, and Baur (1999). It is important to note that according to Fong et al. in order for robots to function in collaborative control, they must be self-reliant, aware, and adaptive. In literature, the adjective "adaptive" is not always shown but is noted in the official sense as it is an important element of collaborative control. The adaptation of traditional applications of control theory in teleoperations sought initially to reduce the sovereignty of "humans as controllers/ robots as tools" and had humans and robots working as peers, collaborating to perform tasks and to achieve common goals. Early implementations of adaptive collaborative control centered on vehicle teleoperation. Recent uses of adaptive collaborative control cover training, analysis, and engineering applications in teleoperationsbetween humans and multiple robots, multiple robots collaborating among themselves, unmanned vehicle control, and fault tolerant controller design.

Like traditional control methodologies, adaptive collaborative control takes inputs into the system and regulates the output based on a predefined set of rules. The difference is that those rules or constraints only apply to the higher-level strategy (goals and tasks) set by humans. Lower tactical level decisions are more adaptive, flexible, and accommodating to varying levels of autonomy, interaction and agent (human and/or robotic) capabilities. Models under this methodology may query sources in the event there is some uncertainty in a task that affects the overarching strategy. That interaction will produce an alternative course of action if it provides more certainty in support of the overarching strategy. If not or there is no response, the model will continue performing as originally anticipated. Several important considerations are necessary for the implementation of adaptive collaborative control for simulation. As discussed earlier, data is provided from multiple collaborators to perform necessary tasks. This basic function requires data fusion on behalf of the model and potentially a need to set a prioritization scheme for handling continuous streaming of recommendations. The degree of autonomy of the robot in the case of human–robot interaction and weighting of decisional authority in robot-robot interaction are important for the control architecture. The design of interfaces is an important human system integration consideration that must be addressed. Due to the inherent varied interpretational scheme in humans, it becomes an important design factor to ensure the robot(s) are correctly conveying its message when interacting with humans.

## Initialization

The simuland for adaptive collaborative control centers on robotics. As such, adaptive collaborative control follows the tenets of control theory applied to robotics at its basest level. That means the states of the robot are observed at a given instant and noted if it is within some accepted bound. If it is not, the estimated states of the robot are calculated using equations of dynamics and kinematics at some future time. The process of entering observation data into the model to generate initial conditions is called initialization. The process of initialization for adaptive collaborative control occurs differently depending on the environment: robotics only and human-robotic interaction. Under a robotics only environment, initialization occurs very similarly to the description above. The robotics, systems, subsystems, non-human entities observe some state it finds not in accordance with the higher-level strategy. The entities that are aware of this error use the appropriate equations to present a revised value for a future time step to its peers. For human-robotic interactions, initialization can occur at two different levels. The first level is what was previously described. In this instance, the robot notices some anomaly in its states that is not wholly consistent or is problematic with its higher-level strategy. It queries the human seeking advice to regulate its dilemma. In the other case, the human feels cause to either query some aspect of the robot's state (e.g. health, trajectory, speed) or present advice to the robot that is challenged against the robot's existing tactical approach to the higher-level strategy. The main inputs for adaptive collaborative control are a human-initiated dialogue based command or value presented by either a human or robotic element. The inputs used in the system models serve as the starting point for the collaboration. A number of ways are available to gather observational data for use in functional models. The easiest method to gather observational data is simple human observation of the robotic system. Self-monitoring attributes such as built-in test (BIT) can provide regular reports on important system characteristics. A common approach to gather observations is to employ sensors throughout the robotic system. Vehicles operating in teleoperations have speedometers to indicate how fast they travel. Robotic systems with either stochastic or cyclic motion often employ accelerometers to note the forces exerted. GPS sensors provide a standardized data type that is used nearly universally for depicting location. Multi-sensor systems have been used to gather heterogeneous observational data for applications in path planning.

## Computation

Adaptive collaborative control is most accurately modeled as a closed loop feedback control system. Closed loop feedback control describes the event where the outputs of a system from an input are used to influence the present or future behavior of the system. The feedback control model is governed by a set of equations that are used to predict the future state of the simuland and regulate its behavior. These equations – in conjunction with principles of control theory – are used to evolve physical operations of the simuland to include, but not limited to: dialogue, path planning, motion, monitoring, and lifting objects over time. Many times, these equations are modeled as nonlinear partial differential equations over a continuous time domain. Due to their complexity, powerful computers are necessary to implement these models. A consequence of using computers to simulate these models is that continuous systems cannot be fully calculated. Instead, numerical solutions, such as the Runge–Kutta methods, are utilized to approximate these continuous models. These equations are initialized from the response of one or more sources and rates of changes and outputs are calculated. These rates of changes predict the states of the simuland a short time in the future. The time increment for this prediction is called a time step. These new states are applied to the model to determine the new rates of changes and observational data. This behavior is continued until the desired number of iterations is completed. In the event a future state violates or comes within a tolerance of the violation the simuland will confer with its human counterpart seeking advice on how to proceed from that point. The outputs, or observational data, are used by the human operators to determine what they believe is the best course of action for the simuland. Their commands are fed with the input into the control system and assessed regarding its effectiveness in resolving the issues. If the human commands are determined to be valuable, the simuland will adjust its control input to what the human suggested. If the human's commands are determined to be unbeneficial, malicious, or non-existent, the model will seek its own correction approach.

## Domain and Codomain

The domain for the models used to conduct adaptive collaborative control is commands, queries, and responses from the human operator at the finite-state machine level. Commands from the human operator allow the agent to be provided with additional input in its decision-making process. This information is particularly beneficial when the human is a subject matter expert or the human is aware of how to reach an overarching goal when the agent is focused on only one aspect of the entire problem. Queries from the human are used to gather status information on either support functions of the agent or to determine progress on missions. Many times the robot's response serves as precursor information for issuance of a command as human assistance to the agent. Responses from the human operator are initiated by queries from the agent and feedback into the system to provide additional input to potentially regulate an action or set of actions from the agent. At the functional model level, the system has translated all accepted commands from the human into control inputs used to carry out tasks defined to the agent. Due to the autonomous nature of the simuland, input from the agent is being fed into the machine to operate sustaining functions and tasking that the human operator has ignored or answered to an insufficient manner. The codomain for the models that utilize adaptive collaborative control are queries, information statements, and responses from the agent. Queries and information statements are elements of the dialogue exchange at the finite-state machine level. Queries from the agent are the system's way of soliciting a response from a human operator. This is particularly important when the agent is physically stuck or at a logical impasse. The types of queries the agent can ask must be pre-defined by the modeler. The frequency and detail associated with a particular query depends on the expertise of the human operator or more accurately the expertise of the human operator identified to the agent. When the agent responds it will send an information statement to the human operator. This statement provides a brief description on what the adaptive collaborative control system decided. At the functional model level, the action associated with the information statement is carried out.

## Applications

### Vehicle Teleoperation

Vehicle teleoperation has been around for many years. Early adaptations of vehicle teleoperations were robotic vehicles that were controlled continuously by human operators. Many of these systems were operated with line-of-sight RF communications and are now regarded as toys for children. Recent developments in the area of unmanned systems have brought a measure of autonomy to the robots. Adaptive collaborative control offers a shared mode of control where robotic vehicles and humans exchange ideas and advice regarding the best decisions to make on a route following and obstacle avoidance. This shared mode of operation mitigates problems of humans remotely operating in hazardous environments with poor communications and limited performance when humans have continuous, direct control. For vehicle teleoperations, robots will query humans to receive input on decisions that affect their tasks or when presented with safety-related issues. This dialogue is presented through an interface module that also allows the human operation to view the impact of the dialogue. In addition, this interface module allows the human operator to view what the robot's sensors capture in order to initiate commands or inquiries as necessary.

## Fault Tolerant System

In practice, there are cases where multiple subsystems work together to achieve a common goal. This is a fairly common practice for reliability engineering. This technique involves systems working together collaboratively and the reliable operation of the overarching system is an important issue. Fault tolerant strategies are combined with the subsystems to form a fault tolerant collaborative system. A direct application is the case where two robotic manipulators work together to grasp a common object. For these systems, it is important that when one subsystem becomes faulty, the healthy subsystem reconfigures itself to operate alone to ensure the whole system can still perform its operations until the other subsystem is repaired. In this case, the subsystems create a dialogue between themselves to determine one another's status. In the event of one system starting to exhibit numerous or dangerous faults the secondary subsystem takes over the operation until the faulty system can be repaired.

## Levels of Autonomy

Four levels of autonomy have been devised to serve as a baseline for human-robot interactions that included adaptive collaborative control. The four levels, ranging from full manual to fully autonomous, are: tele mode, safe mode, shared mode, and autonomous mode. Adaptive collaborative controllers typically range from shared mode to autonomous mode. The two modes of interest are:

- Shared mode – robots can relieve the operator of the burden of direct control, using reactive navigation to find a path based on their perception of the environment. Shared mode provides for a dynamic allocation of roles and responsibilities. The robot accepts varying degrees of operator intervention and supports dialogue through the use of a finite number of scripted suggestions (e.g. "Path blocked! Continue left or right?") and other text messages that appear within the graphical user interface.

- Autonomous mode – robots self-regulate high-level tasks such as patrol, search region or follow path. In this mode, the only user intervention occurs at the tasking level, i.e. the robot manages all decision-making and navigation.

## Limitations

Like many other control strategies, adaptive collaborative control has limits to its capabilities. Although the adaptive collaborative control allows for many tasks to be automated and other predefined cases to query the human operator, unstructured decision making remains the domain of humans, especially when common sense is required. Particularly, robots possess poor judgment at high-level perceptualfunctions, including object recognition and situation assessment. A high number of tasks or a particular task that is very involved may create many questions, thereby increasing the complexity of the dialogue. This complexity to the dialogue in turn adds complexity to the system design. To retain its adaptive nature, the flow of control and information through the simuland will vary with time and events. This dynamic makes debugging, verification, and validation difficult because it is harder to precisely identify an error condition or duplicate a failure situation. This becomes particularly problematic if the system must operate in a regulated facility, such as a nuclear power plant or waste water facility. Issues that affect human-based teams also encumber adaptive collaborative controlled systems. In both cases, teams are required to coordinate activities, exchange information, communicate effectively, and minimize the potential for interference. Other factors that affect teams include resource distribution, timing, sequencing, progress monitoring, and procedure maintenance. Collaboration involves that all partners exhibit trust in the other collaborators and understand the other. To do so, each collaborator needs to have an accurate idea of what the other is capable of doing and how they will carry out an assignment. In some cases, the agent may have to weigh the responses from a human and the human must believe in the decisions a robot makes.

# ROBOT CALIBRATION

Robot calibration is a process used to improve the accuracy of robots, particularly industrial robots which are highly repeatable but not accurate. Robot calibration is the process of identifying certain parameters in the kinematic structure of an industrial robot, such as the relative position of robot links. Depending on the type of errors modeled, the calibration can be classified in three different ways. Level-1 calibration only models differences between actual and reported joint displacement values, (also known as mastering). Level-2 calibration, also known as kinematic calibration, concerns the entire geometric robot calibration which includes angle offsets and joint lengths. Level-3 calibration, also called a non-kinematic calibration, models errors other than geometric defaults such as stiffness, joint compliance, and friction. Often Level-1 and Level-2 calibration are sufficient for most practical needs.

Parametric robot calibration is the process of determining the actual values of kinematic and dynamic parameters of an industrial robot (IR). Kinematic parameters describe the relative position and orientation of links and joints in the robot while the dynamic parameters describe arm and joint masses and internal friction.

Non-parametric robot calibration circumvents the parameter identification. Used with serial robots, it is based on the direct compensation of mapped errors in the workspace. Used with parallel robots, non-parametric calibration can be performed by the transformation of the configuration space.

Robot calibration can remarkably improve the accuracy of robots programmed offline. A calibrated robot has a higher absolute as well as relative positioning accuracy compared to an uncalibrated one; i.e., the real position of the robot end effector corresponds better to the position calculated from the mathematical model of the robot. Absolute positioning accuracy is particularly relevant in connection with robot exchangeability and off-line programming of precision applications. Besides the calibration of the robot, the calibration of its tools and the workpieces it works with (the so-called *cell calibration*) can minimize occurring inaccuracies and improve process security.

## Accuracy Criteria and Error Sources

The international standard ISO 9283 sets different performance criteria for industrial robots and suggests test procedures in order to obtain appropriate parameter values. The most important criteria, and also the most commonly used, are pose accuracy (AP) and pose repeatability (RP). Repeatability is particularly important when the robot is moved towards the command positions manually ( "Teach-In"). If the robot program is generated by a 3D simulation (*off-line programming*), absolute accuracy is vital, too. Both are generally influenced negatively by kinematic factors. Here especially the joint offsets and deviations in lengths and angles between the individual robot links take effect.

## Measurement Systems

There exist different possibilities for pose measurement with industrial robots, e.g. touching reference parts, using supersonic distance sensors, laser interferometry, theodolites, calipers or laser triangulation. Furthermore, there are camera systems which can be attached in the robot's cell or at the IR mounting plate and acquire the pose of a reference object. Measurement and calibration systems are made by such companies as Bluewrist, Dynalog, RoboDK, FARO Technologies, Creaform, Leica, Metris, Metronor, Wiest, Teconsult and Automated Precision,Inc..

## Mathematical Principles

The robot errors gathered by pose measurements can be minimized by numerical optimization. For kinematic calibration, a complete kinematical model of the geometric structure must be developed, whose parameters then can be calculated by mathematical optimization. The common system behaviour can be described with the vector model function as well as input and output vectors. The variables $k, l, m, n$ and their derivates describe the dimensions of the single vector spaces. Minimization of the residual error $r$ for the purpose of identification of the optimal parameter vector $p$ follows from the difference between both output vectors using the Euclidean norm.

For solving the kinematical optimization problems least-squares descent methods are convenient, e.g. a modified quasi-Newton method. This procedure supplies corrected kinematical parameters for the measured machine, which then for example can be used to update the system variables in the controller in order to adapt the used robot model to the real kinematics.

Results:

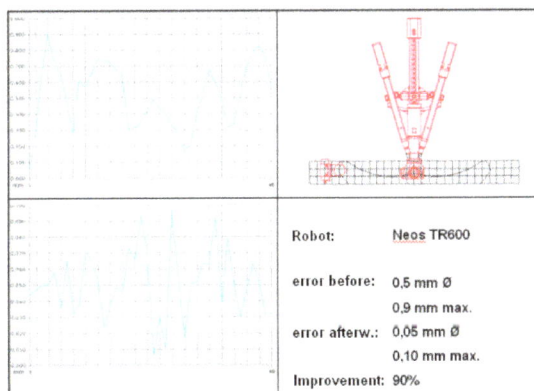

Positioning accuracy of a Tricept robot before and after calibration.

The positioning accuracy of industrial robots varies by manufacturer, age, and robot type. Using kinematic calibration, these errors can be reduced to less than a millimeter in most cases. An example of this is shown in the figure to the right.

Accuracy of 6-axis industrial robots can improved by a factor of 10.

Accuracy of parallel robots after calibration can be as low as a tenth of a millimeter.

## Sample Applications

In-line measurement cell for car body inspection.

In the industry there is a general trend towards substitution of machine tools and special machines by industrial robots for certain manufacturing tasks whose accuracy demands can be fulfilled by calibrated robots. Through simulation and Off-line programming (robotics) it is possible to easily accomplish complex programming tasks, such as robot machining. However, contrary to the teach programming method, good accuracy as well as repeatability is required.

In the figure, a current example is shown: In-line measurement in automotive manufacturing, where the common "measurement tunnel" used for 100% inspection with many expensive sensors are partly replaced by industrial robots that carry only one sensor each. This way the total costs of a measurement cell can be reduced significantly. Furthermore the station can be re-used after a model change by simple re-programming without mechanical adaptations.

Further examples for precision applications are robot-guided hemming in car body manufacturing, assembly of mobile phones, drilling, riveting and milling in aerospace industry and increasingly medical applications.

## References

- Locomotion, definition: techtarget.com, Retrieved 16 August, 2019
- "Robotic fish powered by Gumstix PC and PIC". Human Centred Robotics Group at Essex University. Retrieved 2007-10-25
- Robotlocomotion, classes, hexmoor: cs.siu.edu, Retrieved 17 January, 2019
- M. H. Raibert, H. B. Brown, "Experiments in balance with a 2D one-legged hopping machine," ASME Journal of Dynamic Systems, Measurement, and Control, pp75-81, 1984

- Robot-locomotion-principles: maxembedded.com, Retrieved 18 February 2019

- Wettels, N; Santos, VJ; Johansson, RS; Loeb, Gerald E.; et al. (2008). "Biomimetic tactile sensor array". Advanced Robotics. 22 (8): 829–849. Doi:10.1163/156855308X314533

- Robotlocomotion, classes, hexmoor: cs.siu.edu, Retrieved 19 March, 2019

- Lightcap, C.; Banks, S. (2007-10-01). "Dynamic identification of a mitsubishi pa10-6ce robot using motion capture". 2007 IEEE/RSJ International Conference on Intelligent Robots and Systems: 3860–3865. Doi:10.1109/IROS.2007.4399425. ISBN 978-1-4244-0911-2

# 5

# Applications

Robotics has a wide range of applications in the fields of medicine, agriculture, engineering, navigation and households. Nanorobotics is used for cancer detection and therapy, biohazard defense, etc. This chapter has been carefully written to provide an easy understanding of these applications of robotics.

Robotics is the engineering science and technology which involves the conception, design, operation and manufacture of robots. Electronics, mechanics and software are brought together by robotics.

Robots are used for jobs that are dirty, dull and dangerous. Today robotics have many different application areas. Some of those are:

- Outer Space Applications: Robots are playing a very important role for outer space exploration. The robotic unmanned spacecraft is used as the key of exploring the stars, planets etc.

The most famous robots used in the outer space applications are the Mars rovers of NASA. In 1997 The Pathfinder Mission landed on Mars. Its robotic rover Sojourner, rolled down a ramp and onto Martian soil in early July. It continued to broadcast data from the Martian surface until September.

Sojourner performed semi-autonomous operations on the surface of Mars as part of the Mars Pathfinder mission; equipped with an obstacle avoidance program. Sojourner was capable of planning and navigating routes to study the surface of the planet. Sojourner's ability to navigate with little data about its environment and nearby surroundings allowed the robot to react to unplanned events and objects.

After Sojourner's mission NASA sent twin robots Spirit and Opportunity to the Red Planet on 10, June and 23, July 2003. Spirit and Opportunity landed on Mars on 4, January and 25, January 2004.

Spirit and Opprtunity are solar powered robots with six wheels included their own motors. Both of the Mars Rovers are 1,5 m high, 2,3 m wide and 1,6 m long and weighing 180 kg. Spirit and Opportunity have many science instruments in order to perform their missions on Mars. They have a robot arm, that contains a Mössbauer spectrometer to investigate the mineralogy of the rocks and soils on Mars, an Alpha particle X-ray spectrometer for analysis of elements found in rocks and soils, a rock abrasion tool used to expose the fresh material for examination, a microscopic imager and magnets to collect magnetic particles.

The twin Mars Rovers have a panoramic camera used for examinations of the texture, color, mineralogy, and structure of the local terrain, a miniature thermal emmision spectrometer for identification promising rocks and soils which is useful to determine the formation processes of them. There is also a navigation camera on both Mars rovers in order to take view with a higher field but lower resolution for driving and navigation.

Phoenix Mars Rover was sent to the Red Planet on 4, August 2007 and landed on 25, May 2008. The mission of the Phoenix was to investigate the existence of water and life supporting conditions on Mars.

- Military Applications: In today's modern army robotics is an important factor which is researched and developed day by day. Already remarkable success has been achieved with unmanned aerial vehicles like the Predator drone, which are capable of taking surveillance photographs, and even accurately launching missiles at ground targets, without a pilot. There are many advantages in robotic technology in warfare however, as outlined by Major

Kenneth Rose of the US Army's Training and Doctrine Command: "Machines don't get tired. They don't close their eyes. They don't hide under trees when it rains and they don't talk to their buddies. A human's attention to detail on guard duty drops dramatically in the first 30 minutes. Machines know no fear."

- Intelligent Home Applications: We can monitor home security, environmental conditions and energy usage with intelligent robotic home systems. Door and windows can be opened automatically and appliances such as lighting and air conditioning can be pre programmed to activate. This assists occupants irrespective of their state of mobility.

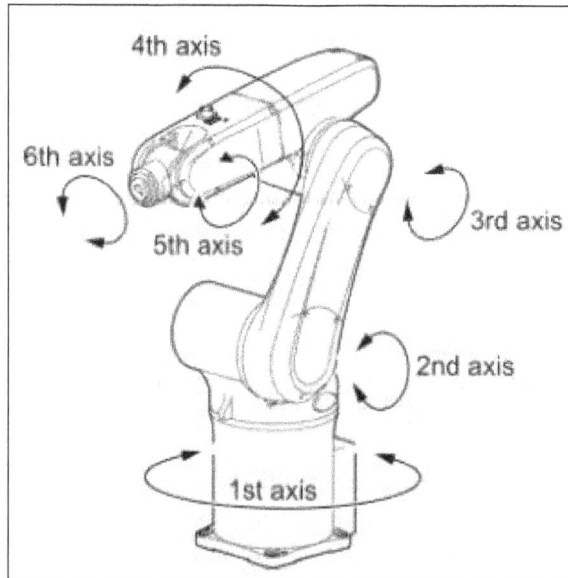

Articulated Robot.

- Industry: From the beginning of the industrial revolution robotics and automation becomes the most important part of manufacturing. Robotic arms which are able to perform multiple tasks such as welding, cutting, lifting, sorting and bending are used in fabrics.

The most commonly used configurations of the industrial robots are:

- Articulated Robots: An articulated robot is one which uses rotary joints to access its work space. Articulated robots can range from simple two-jointed structures to systems with 10 or more interacting joints.The six-axis, articulated robot is the most versatile industrial robot which allows for a high level of freedom.

- Cylindirical Coordinate Robots: These robots have three degrees of freedom and they moves linearly only along the Y and Z axes with a cylindirical work envelope.

- Scara Robots: It stands for Selective Compliant Assembly Robot Arm or Selective Compliant Articulated Robot Arm. SCARA robots usually have four axes as any X-Y-Z coordinate within their work envelope and a fourth axis of motion which is the wrist rotate (Theta-Z).

- Spherical Coordinate Robots: The sperical arm, also known as polar coordinate robot arm, has one sliding motion and two rotational, around the vertical post and around a shoulder joint.

- Cartesian Coordinate Robots: Rectangular arms are sometimes called "Cartesian" because the arm´s axes can be described by using the X, Y, and Z coordinate system. It is claimed that the cartesian design will produce the most accurate movements.

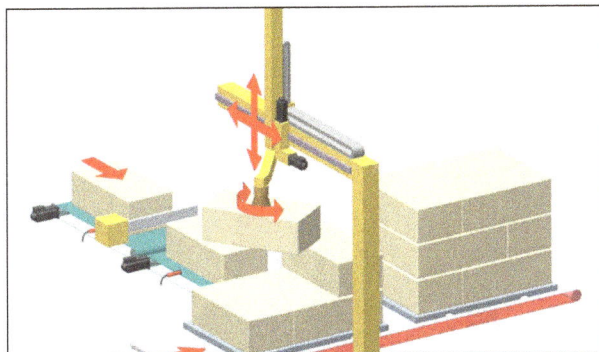

- Delta Robots: A Delta robot consists of three arms connected to universal joints at the base. The key design feature is the use of parallelograms in the arms, which maintains the orientation of the end effector. The Delta robot has popular usage in picking and packaging in factories.

- Health Service: Under development is a robotic suit that will enable nurses to lift patients without damaging their backs. Scientists in Japan have developed a power-assisted suit which will give nurses the extra muscle they need to lift their patients - and avoid back injuries.

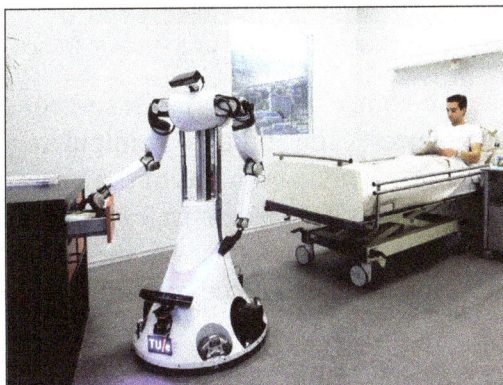

# ROBOTIC APPLICATIONS IN MEDICINE

A wide range of robots is being developed to serve in a variety of roles within the medical environment. Robots specializing in human treatment include surgical robots and rehabilitation robots. The field of assistive and therapeutic robotic devices is also expanding rapidly. These include robots that help patients rehabilitate from serious conditions like strokes, empathic robots that assist in the care of older or physically/mentally challenged individuals, and industrial robots that take on a variety of routine tasks, such as sterilizing rooms and delivering medical supplies and equipment, including medications.

Below are six top uses for robots in the field of medicine today:

1. Telepresence: Physicians use robots to help them examine and treat patients in rural or remote locations, giving them a "telepresence" in the room. "Specialists can be on call, via the robot, to answer questions and guide therapy from remote locations,". "The key features of these robotic devices include navigation capability within the ER, and sophisticated cameras for the physical examination."

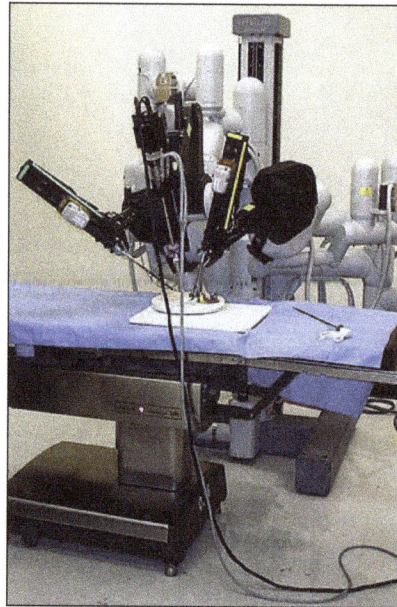

A robotic surgical system controlled by a surgeon from a console.

2. Surgical Assistants: These remote-controlled robots assist surgeons with performing operations, typically minimally invasive procedures. "The ability to manipulate a highly sophisticated robotic arm by operating controls, seated at a workstation out of the operating room, is the hallmark of surgical robots,". Additional applications for these surgical-assistant robots are continually being developed, as more advanced 3DHD technology gives surgeons the spatial references needed for highly complex surgery, including more enhanced natural stereo visualization, combined with augmented reality.

3. Rehabilitation Robots: These play a crucial role in the recovery of people with disabilities, including improved mobility, strength, coordination, and quality of life. These robots can be programmed to adapt to the condition of each patient as they recover from strokes, traumatic brain or spinal cord injuries, or neurobehavioral or neuromuscular diseases such as multiple sclerosis. Virtual reality integrated with rehabilitation robots can also improve balance, walking, and other motor functions.

4. Medical Transportation Robots: Supplies, medications, and meals are delivered to patients and staff by these robots, thereby optimizing communication between doctors, hospital staff members, and patients. "Most of these machines have highly dedicated capabilities for self-navigation throughout the facility". "There is, however, a need for highly advanced and cost-effective indoor navigation systems based on sensor fusion location technology in order to make the navigational capabilities of transportation robots more robust."

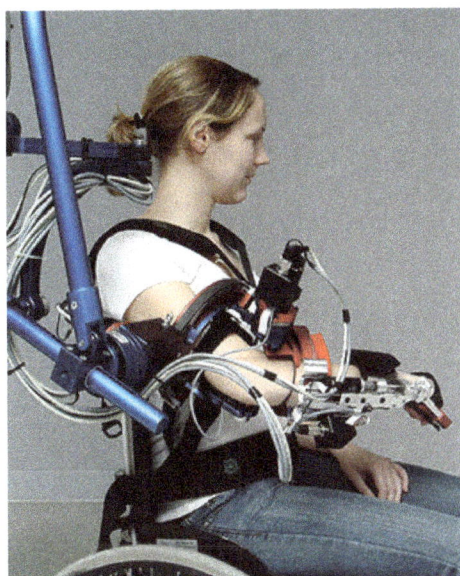

Upper limb rehabilitation.

5. Sanitation and Disinfection Robots: With the increase in antibiotic-resistant bacteria and outbreaks of deadly infections like Ebola, more healthcare facilities are using robots to clean and disinfect surfaces. "Currently, the primary methods used for disinfection are UV light and hydrogen peroxide vapors". "These robots can disinfect a room of any bacteria and viruses within minutes."

6. Robotic Prescription Dispensing Systems: The biggest advantages of robots are speed and accuracy, two features that are very important to pharmacies. "Automated dispensing systems have advanced to the point where robots can now handle powder, liquids, and highly viscous materials, with much higher speed and accuracy than before".

## Future Models

Advanced robots continue to be designed for an ever-expanding range of applications in the healthcare space. A compact, high-precision surgical robot that will operate within the bore of an MRI scanner, as well as the electronic control systems and software that go with it, to improve prostate biopsy accuracy.

Since the MRI scanner uses a powerful magnet, the robot, including all of its sensors and actuators, must be made from nonferrous materials. "On top of all this, we had to develop the communications protocols and software interfaces for controlling the robot, and interface those with higher-level imaging and planning systems," says Fischer. "The robot must be easy for a non-technical surgical team to sterilize, set up, and place in the scanner. This all added up to a massive systems integration project which required many iterations of the hardware and software to get to that point."

In other research, virtual reality is being integrated with rehabilitation robots to expand the range of therapy exercise, increasing motivation and physical treatment effects. Exciting discoveries are being made with nanoparticles and nanomaterials. For example, nanoparticles can traverse the "blood-brain barrier." In the future, nanodevices can be loaded with "treatment payloads" of medicine that can be injected into the body and automatically guided to the precise target sites within the body. Soon, ingestible, broadband-enabled digital tools will be available that use wireless technology to help monitor internal reactions to medications.

"Existing technologies are being combined in new ways to streamline the efficiency of healthcare operations". "While at the same time, emerging robotic technologies are being harnessed to enable intriguing breakthroughs in medical care."

# ROBOT-ASSISTED SURGERY

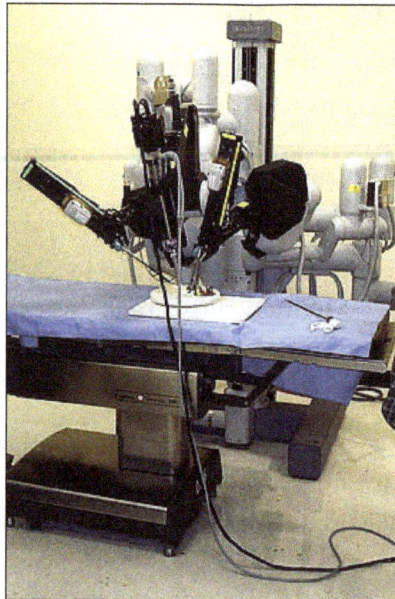

A robotically assisted surgical system used for prostatectomies, cardiac valve repair and gynecologic surgical procedures.

Robotic surgery are types of surgical procedures that are done using robotic systems. Robotically-assisted surgery was developed to try to overcome the limitations of pre-existing minimally-invasive surgical procedures and to enhance the capabilities of surgeons performing open surgery.

In the case of robotically-assisted minimally-invasive surgery, instead of directly moving the instruments, the surgeon uses one of two methods to administer the instruments. These include using a direct telemanipulator or through computer control. A telemanipulator is a remote manipulator that allows the surgeon to perform the normal movements associated with the surgery. The robotic arms carry out those movements using end-effectors and manipulators to perform the actual surgery. In computer-controlled systems, the surgeon uses a computer to control the robotic arms and its end-effectors, though these systems can also still use telemanipulators for their input. One advantage of using the computerized method is that the surgeon does not have to be present, leading to the possibility for remote surgery.

Robotic surgery has been criticized for its expense, with the average costs in 2007 ranging from $5,607 to $45,914 per patient. This technique has not been approved for cancer surgery as of 2019 as the safety and usefulness is unclear.

# Uses

## Heart

As of 2004, three types of heart surgery are being performed using robotic surgery systems:

- Atrial septal defect repair – the repair of a hole between the two upper chambers of the heart.

- Mitral valve repair – the repair of the valve that prevents blood from regurgitating back into the upper heart chambers during contractions of the heart.

- Coronary artery bypass – rerouting of blood supply by bypassing blocked arteries that provide blood to the heart.

## Thoracic

Robotic surgery has become more widespread in thoracic surgery for mediastinal pathologies, pulmonary pathologies and more recently complex esophageal surgery.

## Gastrointestinal

Multiple types of procedures have been performed with either the 'Zeus' or da Vinci robot systems, including bariatric surgery and gastrectomy for cancer. Surgeons at various universities initially published case series demonstrating different techniques and the feasibility of GI surgery using the robotic devices. Specific procedures have been more fully evaluated, specifically esophageal fundoplication for the treatment of gastroesophageal reflux and Heller myotomy for the treatment of achalasia.

Robot-assisted pancreatectomies have been found to be associated with "longer operating time, lower estimated blood loss, a higher spleen-preservation rate, and shorter hospital stay[s]" than laparoscopic pancreatectomies; there was "no significant difference in transfusion, conversion to open surgery, overall complications, severe complications, pancreatic fistula, severe pancreatic fistula, ICU stay, total cost, and 30-day mortality between the two groups."

## Gynecology

Robotic surgery in gynecology is of uncertain benefit as of 2019 with it being unclear if it affects rates of complications. Gynecologic procedures may take longer with robot-assisted surgery but may be associated with a shorter hospital stay following hysterectomy. In the United States, robotic-assisted hysterectomy for benign conditions has been shown to be more expensive than conventional laparoscopic hysterectomy, with no difference in overall rates of complications.

This includes the use of the da Vinci surgical system in benign gynecology and gynecologic oncology. Robotic surgery can be used to treat fibroids, abnormal periods, endometriosis, ovarian tumors, uterine prolapse, and female cancers. Using the robotic system, gynecologists can perform hysterectomies, myomectomies, and lymph node biopsies. The Memic robotic system is aimed to provide a robotic platform for natural orifice transluminal endoscopic surgery (NOTES) for myomectomy through the vagina.

A 2017 review of surgical removal of the uterus and cervix for early cervical cancer robotic and laparoscopic surgery resulted in similar outcomes with respect to the cancer.

## Bone

Robots are used in orthopedic surgery.

## Spine

Robotic devices started to be used in minimally invasive spine surgery starting in the mid-2000s. As of 2014, there were too few randomized clinical trials to judge whether robotic spine surgery is more or less safe than other approaches.

As of 2019, the application of robotics in spine surgery has mainly been limited to pedicle screw insertion for spinal fixation. In addition, the majority of studies on robot-assisted spine surgery have investigated lumbar or lumbosacral vertebrae only. Studies on use of robotics for placing screws in the cervical and thoracic vertebrae are limited.

## Transplant Surgery

The first fully robotic kidney transplantations were performed in the late 2000s. It may allow kidney transplantations in people who are obese who could not otherwise have the procedure. Weight loss however is the preferred initial effort.

## General Surgery

With regards to robotic surgery, this type of procedure is currently best suited for single-quadrant procedures, in which the operations can be performed on any one of the four quadrants of the abdomen. Cost disadvantages are applied with procedures such as a cholecystectomy and fundoplication, but are suitable opportunities for surgeons to advance their robotic surgery skills.

## Urology

Robotic surgery in the field of urology has become common, especially in the United States. There is inconsistent evidence of increased benefits compared to standard surgery to justify the increased costs. Some have found tentative evidence of more complete removal of cancer and fewer side effects from surgery for prostatectomy. In 2000, the first robot-assisted laparoscopic radical prostatectomy was performed.

## Underwater Robotics

This area deals with the development and realization of Artificial Intelligence methods in underwater systems. Main points of research are:

- Development of systems for user support in remote-controlled underwater vehicles employing virtual immersion methods.

- Design of methods for autonomous manipulation and mission planning of robot arms in underwater applications, particularly with state-of-the-art sensor technology, such as "Visual Servoing".

- Image evaluation and object recognition with modular and intelligent underwater cameras.

- Design of control methods for next-generation autonomous underwater vehicles.

- Development of biologically inspired and energy-efficient methods of transport for underwater vehicles, such as oscillating systems.

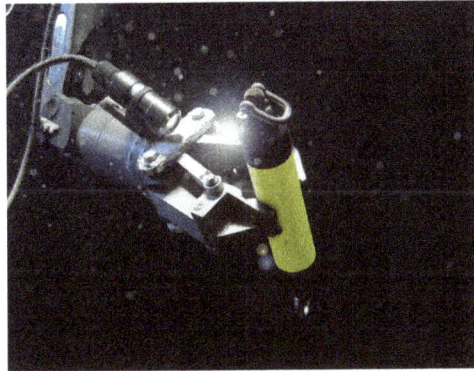
Underwater Robotics: CManipulator.

## Electric Mobility

Electric Mobility: EO2 smart connecting car.

In the field of electric mobility we are testing concepts for electric vehicles, battery charge technologies, and the collection of vehicle data. We are creating models for intelligent, environmentally sound, and integrated urban mobility. Our research focuses around:

- Development and demonstration of innovative vehicle concepts.

- Design of new approaches to mobility and traffic control, application support, technology integration.

- Data collection by fleet tests with technologically different electric vehicles.

- Coordination of the regional project office of the model region Electric Mobility Bremen/ Oldenburg.

- Virtualization of the model region, simulation of future, larger vehicle fleets, and predictions of the effects on the model region in terms of traffic volume, infrastructure needs, environmental pollution, and economic efficiency.

- Creating a foundation for new business models and traffic concepts on the basis of the data previously collected.

## Logistics, Production and Consumer (LPC)

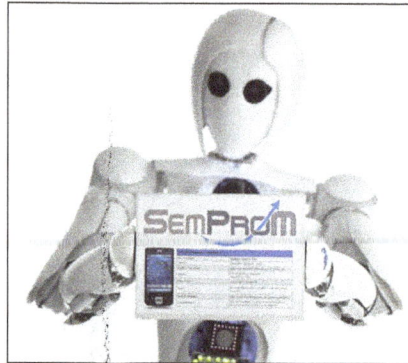

LPC: Robotlady AILA.

In this area, robots are developed to act autonomously and/or support humans in intralogistic, industrial and consumer scenarios. Our research focuses around the new robotics for the Industrie 4.0 and beyond:

- Intelligent human-robot collaboration using hybrid teams for production environments.

- Development of cognitively enhanced robot capabilities for flexible manufacturing.

- Modular, novel and safe robots for human-robot collaboration.

- Autonomous mobile manipulation for intralogistics and manufacturing scenarios.

- Innovative robotics solutions for inspection (for instance, ballast water tanks, ship structures, or tunnel boring machines).

## Search and Rescue (SAR) and Security Robotics

SAR: Advanced Security Guard.

In this area, robots will be developed to support rescue and security personnel. Main points of our research are:

- Development of highly mobile platforms for indoor and outdoor applications.

- Development of autonomous systems that are able to identify potential victims (SAR) or intruders (Security).

- Development and application of state-of-the-art sensor technology based on radar, laser scanner, and thermal vision to identify objects and persons, resp.

- Embedding of robot systems into existing rescue and security infrastructures.

- Autonomous navigation and mission planning.

## Assistance and Rehabilitation Systems

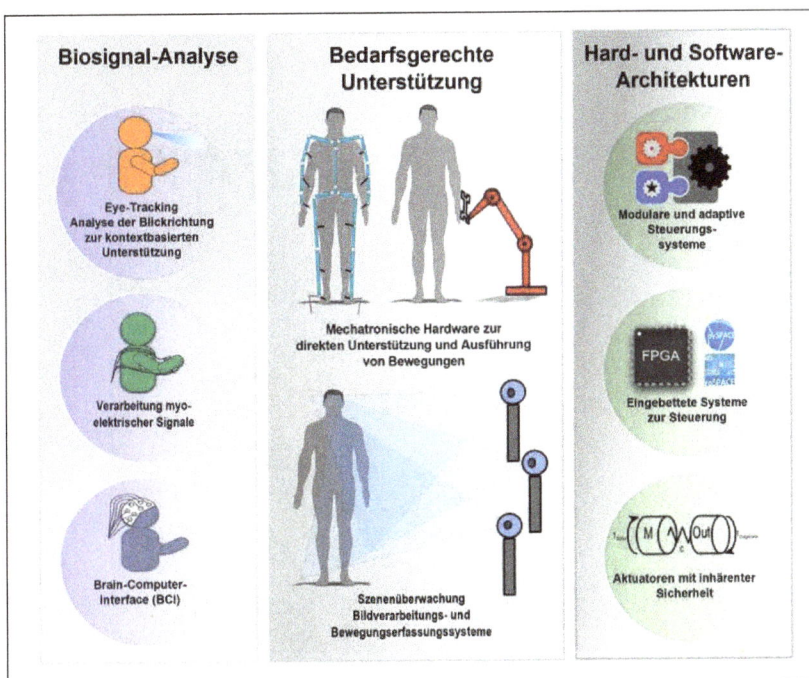

Representation of various components within the scope of Assistance and Rehabilitation Systems.

This field deals with robotic systems that can support humans in complex, exhausting or often repeated tasks. Application areas are both help during activites of everyday life (at home or work) and medical rehabilitation. Support can either take place using systems the human is wearing like exoskeletons or orthoses, or by service robots performing the respective task.

## Core Topics Include

- Concept development, design and construction.

- Intelligent hardware-system architectures.

- Software architectures.

- Embedded biosignal analysis, e.g. using information from:

    ◦ Muscle (EMG).

    ◦ Eye (eyetracking, EOG).

    ◦ Or from brain activity (EEG).

- Fusion of different sensors.

- Direct online signal processing (hard- and software).

- Robust learning systems capable to adapt.

- Joint communication layers for better human-machine interaction.

- (Semi-)autonomously acting systems.

- Assist-as-needed.

## Agricultural Robotics

Cooperative threshing with several machines.

We develop robots for agricultural applications and transfer methods and algorithms from robotics to conventional agricultural machines. Our objective is to increase the performance of machines and processes and to reduce resource consumption at the same time. Our research is focused on technology applications used in the cultivation of land. Primary research topics are:

- Methods for autonomous planning and navigation of outdoor machinery.

- Methods for environmental recognition in agricultural machinery control.

- Methods of infield logistics to optimize cooperation and resource consumption between multiple agricultural machines.

- Interoperability at the level of communication, processes and knowledge processing.

# ROBOTIC APPLICATIONS IN AGRICULTURAL INDUSTRY

The robots can be used on farms and they enable precision agriculture techniques, they are used to autonomously monitor soil respiration, photosynthetic activity, leaf area indexes (LAI) and other biological factors, The robots are used to herd livestock on large ranches, They monitor the animals and ensure they are healthy and have enough area to graze.

The robots are equipped to monitor the pollution created by agriculture at the ground level, They can measure carbon dioxide & nitrous oxide emissions, so, the farmers can reduce their environmental footprint, Crop Harvesting robots can work around the clock for faster harvesting, in some cases completing the same amount of work as approximately 30 workers.

Robots offer an efficient method in spraying pesticides, Micro-spraying robots can autonomously navigate the farm and deliver targeted sprays of herbicides to eliminate weeds, This approach reduces crops' exposure to herbicides and helps prevent the growth of herbicide-resistant weeds.

Fruit Harvesting robots are used to harvest fruit in addition to crops, Fruit harvesting is notoriously difficult for robots, These robots are equipped with the advanced vision systems to identify fruits and they can grasp them without damaging them, The robots with 3D vision systems can accurately plant & seed crops for optimal growth.

Nursery Automation robots can create major efficiencies for crop nurseries, primarily to move plants around large greenhouses, they help address a growing labor shortage, Nursery automation is where seeds are grown into young plants, which are later planted outside, Nursery plants are sold directly to consumers and landscape gardeners.

The robots can be attachments to a tractor, As humans drive the tractors, the robots are designed to adapt to the speed that the human is driving, However, fully-autonomous tractors are also becoming popular, However most agricultural robots are applied in crop growing, there have been a few emerging applications within sheep & cattle farming, Collaborative robots are used to help in the milking process on dairy farms, The UR5 can be used to spray disinfectant on the cow's udders in preparation for milking.

Robots are used in crop seeding, Autonomous precision seeding combines robotics with geomapping, The map is generated which shows the soil properties (quality, density, etc) at every point in the field, The tractor, with robotic seeding attachment, places the seeds at precise locations & depths so that each has the best chance of growing.

Robots are used in crop monitoring and analysis, New sensor and geomapping technologies enable the farmers to get a much higher level of data about their crops than they have in the past, Ground robots & drones provide a way to collect this data autonomously.

The farmer can move the drone to the field, and initiate the software via a tablet or smartphone, and view the collected crop data in real time, Ground-based robots provide more detailed monitoring as they can get closer to the crops, Some can be used for other tasks like weeding and fertilizing.

Robot-Assisted Precision Irrigation can reduce wasted water by targeting specific plants, Ground robots autonomously navigate between rows of crop and pour water directly at the base of each plant, Robots are used in picking & harvesting crops, It can be done with a combine harvester, which can be automated just like a tractor.

Robots can access areas where other machines can't, Robots present a higher quality of fresh products, lower production costs, and a decreased need for manual labor, They can be used to automate manual tasks, Thinning robot uses computer vision to detect plants as it drives over them and decides which plants to keep and which to remove, Thinning reduces the density of plants so that each has a better chance of growing, Pruning involves cutting back parts of plants to improve their growth.

Weeding robots are used in weeding the plants, They use computer vision to detect the plants as it is pushed by the tractor, It automatically hoes the spaces between the plants to uproot the weeds, Other weeding robots use lasers to kill the weeds and they don't need to use chemicals.

Farmers can use fruit picking robots, driverless tractor/sprayers, and sheep shearing robots, They are designed to replace human labor, Robots can be used in pruning, weeding, spraying and monitoring, They can be used in livestock applications (livestock robotics) such as automatic milking, washing and castrating.

# HUMAN INTERACTION WITH ROBOTS

Human-Robot Interaction (HRI) is a field of study dedicated to understanding, designing, and evaluating robotic systems for use by or with humans. Interaction, by definition, requires communication between robots and humans. Communication between a human and a robot may take several forms, but these forms are largely influenced by whether the human and the robot are in close proximity to each other or not. Thus, communication and, therefore, interaction can be separated into two general categories:

- Remote interaction – The human and the robot are not co-located and are separated spatially or even temporally (for example, the Mars Rovers are separated from earth both in space and time).

- Proximate interactions – The humans and the robots are co-located (for example, service robots may be in the same room as humans).

Within these general categories, it is useful to distinguish between applications that require mobility, physical manipulation, or social interaction. Remote interaction with mobile robots often is referred to as teleoperation or supervisory control, and remote interaction with a physical manipulator is often referred to as telemanipulation. Proximate interaction with mobile robots may take the form of a robot assistant, and proximate interaction may include a physical interaction. Social interaction includes social, emotive, and cognitive aspects of interaction. In social interaction, the humans and robots interact as peers or companions.

## The Goal of Friendly Human–robot Interactions

Robots are artificial agents with capacities of perception and action in the physical world often referred by researchers as workspace. Their use has been generalized in factories but nowadays they tend to be found in the most technologically advanced societies in such critical domains as search and rescue, military battle, mine and bomb detection, scientific exploration, law enforcement, entertainment and hospital care.

Kismet can produce a range of facial expressions.

These new domains of applications imply a closer interaction with the user. The concept of closeness is to be taken in its full meaning, robots and humans share the workspace but also share goals in terms of task achievement. This close interaction needs new theoretical models, on one hand for the robotics scientists who work to improve the robots utility and on the other hand to evaluate the risks and benefits of this new "friend" for our modern society.

With the advance in AI, the research is focusing on one part towards the safest physical interaction but also on a socially correct interaction, dependent on cultural criteria. The goal is to build an intuitive, and easy communication with the robot through speech, gestures, and facial expressions.

Dautenhahn refers to friendly Human–robot interaction as "Robotiquette" defining it as the "social rules for robot behaviour (a 'robotiquette') that is comfortable and acceptable to humans" The robot has to adapt itself to our way of expressing desires and orders and not the contrary. But every day environments such as homes have much more complex social rules than those implied by factories or even military environments. Thus, the robot needs perceiving and understanding capacities to build dynamic models of its surroundings. It needs to categorize objects, recognize and locate humans and further their emotions. The need for dynamic capacities pushes forward every sub-field of robotics.

Furthermore, by understanding and perceiving social cues, robots can enable collaborative scenarios with humans. For example, with the rapid rise of personal fabrication machines such as desktop 3d printers, laser cutters, etc., entering our homes, scenarios may arise where robots can collaboratively share control, co-ordinate and achieve tasks together. Industrial robots have already been integrated into industrial assembly lines and are collaboratively working with humans. The social impact of such robots have been studied and has indicated that workers still treat robots and social entities, rely on social cues to understand and work together.

On the other end of HRI research the cognitive modelling of the "relationship" between human and the robots benefits the psychologists and robotic researchers the user study are often of interests on both sides. This research endeavours part of human society. For effective *human – humanoid robot* interaction numerous communication skills and related features should be implemented in the design of such artificial agents/systems.

## General HRI Research

HRI research spans a wide range of fields, some general to the nature of HRI.

## Methods for Perceiving Humans

Methods for perceiving humans in the environment are based on sensor information. Research on sensing components and software led by Microsoft provide useful results for extracting the human kinematics. An example of older technique is to use colour information for example the fact that for light skinned people the hands are lighter than the clothes worn. In any case a human modelled a priori can then be fitted to the sensor data. The robot builds or has (depending on the level of autonomy the robot has) a 3D mapping of its surroundings to which is assigned the humans locations.

Most methods intend to build a 3D model through vision of the environment. The proprioception sensors permit the robot to have information over its own state. This information is relative to a reference.

A speech recognition system is used to interpret human desires or commands. By combining the information inferred by proprioception, sensor and speech the human position and state (standing, seated). In this matter, Natural language processing is concerned with the interactions between computers and human (natural) languages, in particular how to program computers to process and analyze large amounts of natural language data. For instance, neural network architectures and learning algorithms that can be applied to various natural language processing tasks including part-of-speech tagging, chunking, named entity recognition, and semantic role labeling.

## Methods for Motion Planning

Motion planning in dynamic environment is a challenge that is for the moment only achieved for 3 to 10 degrees of freedom robots. Humanoid robots or even 2 armed robots that can have up to 40 degrees of freedom are unsuited for dynamic environments with today's technology. However lower-dimensional robots can use potential field method to compute trajectories avoiding collisions with human.

## Cognitive Models and Theory of Mind

Humans exhibit negative social and emotional responses as well as decreased trust toward some robots that closely, but imperfectly, resemble humans; this phenomenon has been termed the "Uncanny Valley." However recent research in telepresence robots has established that mimicking human body postures and expressive gestures has made the robots likeable and engaging in a remote setting. Further, the presence of a human operator was felt more strongly when tested with an android or humanoid telepresence robot than with normal video communication through a monitor.

While there is a growing body of research about users' perceptions and emotions towards robots, we are still far from a complete understanding. Only additional experiments will determine a more precise model.

Based on past research, we have some indications about current user sentiment and behavior around robots:

- During initial interactions, people are more uncertain, anticipate less social presence, and have fewer positive feelings when thinking about interacting with robots, and prefer to communicate with a human. This finding has been called the human-to-human interaction script.

- It has been observed that when the robot performs a proactive behaviour and does not respect a "safety distance" (by penetrating the user space) the user sometimes expresses fear. This fear response is person-dependent.

- It has also been shown that when a robot has no particular use, negative feelings are often expressed. The robot is perceived as useless and its presence becomes annoying.

- People have also been shown to attribute personality characteristics to the robot that were not implemented in software.

## Methods for Human-robot Coordination

A large body of work in the field of human-robot interaction has looked at how humans and robots may better collaborate. The primary social cue for humans while collaborating is the shared perception of an activity, to this end researchers have investigated anticipatory robot control through various methods including: monitoring the behaviors of human partners using eye tracking, making inferences about human task intent, and proactive action on the part of the robot. The studies revealed that the anticipatory control helped users perform tasks faster than with reactive control alone.

A common approach to program social cues into robots is to first study human-human behaviors and then transfer the learning. For example, coordination mechanisms in human-robot collaboration are based on work in neuroscience which examined how to enable joint action in human-human configuration by studying perception and action in a social context rather than in isolation. These studies have revealed that maintaining a shared representation of the task is crucial for accomplishing tasks in groups. For example, the authors have examined the task of driving together by separating responsibilities of acceleration and braking i.e., one person is responsible for accelerating and the other for braking; the study revealed that pairs reached the same level of performance as individuals only when they received feedback about the timing of each other's actions. Similarly, researchers have studied the aspect of human-human handovers with household scenarios like passing dining plates in order to enable an adaptive control of the same in human-robot handovers. Most recently, researchers have studied a system that automatically distributes assembly tasks among co-located workers to improve co-ordination.

# ROBOT NAVIGATION

For any mobile device, the ability to navigate in its environment is important. Avoiding dangerous situations such as collisions and unsafe conditions (temperature, radiation, exposure to weather, etc.) comes first, but if the robot has a purpose that relates to specific places in the robot environment, it must find those places. This topic will present an overview of the skill of navigation and try to identify the basic blocks of a robot navigation system, types of navigation systems, and closer look at its related building components.

Robot navigation means the robot's ability to determine its own position in its frame of reference and then to plan a path towards some goal location. In order to navigate in its environment, the robot or any other mobility device requires representation, i.e. a map of the environment and the ability to *interpret* that representation.

Navigation can be defined as the combination of the three fundamental competences:

- Self-localisation.

- Path planning.

- Map-building and map interpretation.

Some robot navigation systems are capable of simultaneous localization and mapping based on 3D reconstructions of their surroundings.

Robot localization denotes the robot's ability to establish its own position and orientation within the frame of reference. Path planning is effectively an extension of localisation, in that it requires the determination of the robot's current position and a position of a goal location, both within the same frame of reference or coordinates. Map building can be in the shape of a metric map or any notation describing locations in the robot frame of reference.

## Vision-based Navigation

Vision-based navigation or optical navigation uses computer vision algorithms and optical sensors, including laser-based range finder and photometric cameras using CCD arrays, to extract the visual features required to the localization in the surrounding environment. However, there are a range of techniques for navigation and localization using vision information, the main components of each technique are:

- Representations of the environment.

- Sensing models.

- Localization algorithms.

In order to give an overview of vision-based navigation and its techniques, we classify these techniques under indoor navigation and outdoor navigation.

## Indoor Navigation

The easiest way of making a robot go to a goal location is simply to *guide* it to this location. This guidance can be done in different ways: burying an inductive loop or magnets in the floor, painting lines on the floor, or by placing beacons, markers, bar codes etc. in the environment. Such *Automated Guided Vehicles* (AGVs) are used in industrial scenarios for transportation tasks. Indoor Navigation of Robots are possible by IMU based indoor positioning devices.

There are a very wider variety of indoor navigation systems. The basic reference of indoor and outdoor navigation systems is "Vision for mobile robot navigation: a survey" by Guilherme N. De-Souza and Avinash C. Kak.

## Outdoor Navigation

Some recent outdoor navigation algorithms are based on convolutional neural network and machine learning, and are capable of accurate turn-by-turn inference.

## Autonomous Flight Controllers

Typical Open Source Autonomous Flight Controllers have the ability to fly in full automatic mode and perform the following operations;

- Take off from the ground and fly to a defined altitude.

- Fly to one or more waypoints.

- Orbit around a designated point.

- Return to the launch position.

- Descend at a specified speed and land the aircraft.

The onboard flight controller relies on GPS for navigation and stabilized flight, and often employ additional Satellite-based augmentation systems (SBAS) and altitude (barometric pressure) sensor.

# DIFFERENT PURPOSES OF ROBOTS

Working deep under water, exploring live volcanoes at short distances or travelling to faraway planets are simply impossible for humans to execute. Robots are often called upon to perform underwater salvage missions to find sunken ships or planes. In 1985, a team of researchers and robot called Jason Junior were able to locate the wreck of the Titanic. Underwater robots operating one and half kilometres under the ocean's surface played a vital role in the fight to stop oil gushing into the Gulf of Mexico.

In manufacturing, many jobs in factories are messy or dirty. Dirty tasks may include welding, grinding, molding and casting. When robots are used to perform these tasks, it enables human workers to partake in more meaningful and creative pursuits.

In manufacturing, robots often perform tasks which are very dangerous for people. Using robots for tasks involving extreme temperatures, for example, reduces the risk of workplace accidents. Apart from dangerous tasks in manufacturing, robots are also being used to carry out other important but dangerous activities such as clearing landmines, helping in rescue missions and mopping up toxic leaks. Police robots are used to defuse and remove explosive devices. Sometimes, police may have to detonate the device on-site. Some robots are so tough they can survive multiple blasts.

# APPLICATION OF NANOROBOTICS

Nanorobotics describes the technology of producing machines or robots at the nanoscale. 'Nanobot' is an informal term to refer to engineered nano machines. Though currently hypothetical, nanorobots will advance many fields through the manipulation of nano-sized objects.

The field of medicine is expected to receive the largest improvement from this technology. This is because nanotechnology provides the advantage of transporting large amounts of nanorobots in a single injection. Furthermore, designs that include a communication interface will allow adaptations to the programming and function of nanobots already in the body. This will improve disease monitoring and treatment whilst reducing the need for invasive procedures.

## Nanorobotic Applications in the Field of Hematology

Current research is developing nanorobotic applications for the field of hematology. This ranges from developing artificial methods of transporting oxygen in the body after major trauma to forming improved clotting capabilities in the event of a dangerous hemorrhage.

Respirocytes are hypothetical nanobots engineered to function as artificial red blood cells. In emergencies where a patient stops breathing and blood circulation ceases, respirocytes could be injected into the blood stream to transport respiratory gases until the patient is stabilized.

Current proposals suggest respirocytes would be able to supply 200 times more respiratory gas molecules than natural red blood cells of the same volume. Clottocytes are another type of nanobot which function as artificial platelets for halting bleeds.

Clottocytes would mimic the natural platelet ability to accumulate at the bleed, in order to form a barrier, by unfurling a fiber mesh which would trap blood cells when the nanobot arrives at the site of the injury. The clotting ability of one injection of clottocytes would be 10,000 times more effective that an equal volume of natural platelets.

## Nanorobotics Applications for Cancer Detection and Therapy

As cancer survival rates improve with early detection, nanorobots designed with enhanced detection abilities will be able to increase the speed of a cancer diagnosis and therefore enhance the prognosis of the disease. Nanobots with embedded chemical sensors can be designed to detect tumor cells in the body. Proposed designs currently include the employment of integrated communication technology, where two-way signaling is produced. This means that nanobots will respond to acoustic signals and receive programming instructions via external sound waves along with transmitting data they have accumulated.

A simple reporting interface could be produced through strategically positioned nanobots in the body which are able to log information supplied by active nanobots traveling through the blood stream. Instructions could be adapted in vivo to provide active targeting for monitoring or healing.

Nanorobots with chemical sensors can also be utilized for therapy. Through specific programming to detect different levels of cancer biomarkers such as e-cadherins and beta-catenin, therapy can be provided in both primary and metastatic phases of cancer. Nanobots have the advantage of producing targeted treatment. Current cancer treatments have severe side effects caused by the destruction of healthy cells. Targeted treatment can be formed by designing nanorobots with chemotactic sensors on their surface which correspond to specific antigens on the cancer cells.

## Nanorobotics Applications for Biohazard Defense

Nanorobots will also have useful applications for biohazard defense, including improving the response to epidemic disease. Nanobots with protein based biosensors will be able to transmit real-time information in areas where public infrastructure is limited and laboratory analysis is unavailable. This is particularly applicable for biomedical monitoring of areas devastated by epidemic disease as well as in remote or war torn countries during humanitarian missions.

Nanorobotics may also reduce contamination and provide successful screening for quarantine. In the event of an influenza epidemic for example, increased concentrations of alpha-NAGA enzyme in the blood stream could be used as a biomarker for the influenza infection. The increased concentration would trigger the nanorobot prognostic protocol which sends electromagnetic back propagated signals to portable technology such as a mobile phone. The information would then be retransmitted via the telecommunication system providing information on the location of the infected person, increasing the speed of contamination quarantine.

## References

- Robotics-applications: robotiksistem.com, Retrieved 20 April, 2019

- Digioia AM, Jaramaz B, Picard F, Nolte L, eds. (30 December 2004). Computer and robotic assisted hip and knee surgery. Oxford University Press. Pp. 127–156. ISBN 978-0-19-850943-1

- Top-6-robotic-applications-in-medicine, content, topics-resources: asme.org, Retrieved 21 May, 2019

- Nathan, Joseph H.; Shvalb, Nir; Smorgick, Noam (2016). "Robotic-Assisted Laparoscopic Myomectomy versus Traditional Laparoscopic Myomectomy: Are They the Same?". Current Obstetrics and Gynecology Reports. 5 (4): 341–347. Doi:10.1007/s13669-016-0182-y. ISSN 2161-3303

- Fields-of-application, research: robotik.dfki-bremen.de, Retrieved 22 June, 2019

- "New Robot Technology Eases Kidney Transplants". CBS News. 22 June 2009. Retrieved 8 July 2009

- Robotic-applications-in-agricultural-industry-autonomous-agricultural-robot-types-uses-and-importance, robotics: online-sciences.com, Retrieved 23 July, 2019

- Kozima, Hideki; Michalowski, Marek P.; Nakagawa, Cocoro (19 November 2008). "Keepon". International Journal of Social Robotics. 1 (1): 3–18. doi:10.1007/s12369-008-0009-8

- 1-introduction: humanrobotinteraction.org, Retrieved 24 August, 2019

- Adalgeirsson, Sigurdur; Breazeal, Cynthia (2010). MeBot: A Robotic Platform for Socially Embodied Presence (pdf). Hri '10. pp. 15–22. ISBN 9781424448937

# Permissions

All chapters in this book are published with permission under the Creative Commons Attribution Share Alike License or equivalent. Every chapter published in this book has been scrutinized by our experts. Their significance has been extensively debated. The topics covered herein carry significant information for a comprehensive understanding. They may even be implemented as practical applications or may be referred to as a beginning point for further studies.

We would like to thank the editorial team for lending their expertise to make the book truly unique. They have played a crucial role in the development of this book. Without their invaluable contributions this book wouldn't have been possible. They have made vital efforts to compile up to date information on the varied aspects of this subject to make this book a valuable addition to the collection of many professionals and students.

This book was conceptualized with the vision of imparting up-to-date and integrated information in this field. To ensure the same, a matchless editorial board was set up. Every individual on the board went through rigorous rounds of assessment to prove their worth. After which they invested a large part of their time researching and compiling the most relevant data for our readers.

The editorial board has been involved in producing this book since its inception. They have spent rigorous hours researching and exploring the diverse topics which have resulted in the successful publishing of this book. They have passed on their knowledge of decades through this book. To expedite this challenging task, the publisher supported the team at every step. A small team of assistant editors was also appointed to further simplify the editing procedure and attain best results for the readers.

Apart from the editorial board, the designing team has also invested a significant amount of their time in understanding the subject and creating the most relevant covers. They scrutinized every image to scout for the most suitable representation of the subject and create an appropriate cover for the book.

The publishing team has been an ardent support to the editorial, designing and production team. Their endless efforts to recruit the best for this project, has resulted in the accomplishment of this book. They are a veteran in the field of academics and their pool of knowledge is as vast as their experience in printing. Their expertise and guidance has proved useful at every step. Their uncompromising quality standards have made this book an exceptional effort. Their encouragement from time to time has been an inspiration for everyone.

The publisher and the editorial board hope that this book will prove to be a valuable piece of knowledge for students, practitioners and scholars across the globe.

# Index

**A**

Accelerometers, 68, 74, 158, 203

Aerobot, 39, 49-51

Agricultural Robot, 39, 53-54

Artificial Intelligence, 26, 30, 35-36, 39-40, 57, 63-64, 74, 79, 82, 137, 188, 218

Automated Guided Vehicles, 41, 66, 228

Autonomous Robot, 39, 46, 57-58, 62, 67, 81

Autonomous Underwater Vehicles, 65, 219

**B**

Beam Robotics, 1, 7-8

Biomimetic Robots, 46, 177

Brain-based Device, 14, 16-18

**C**

Calibration Algorithm, 99

Capacitive Sensor, 150-151

Cartesian Coordinate System, 84, 165

Cartesian Robot, 84, 138

Cloud Computing, 11, 13

Cloud Robotics, 11, 13

Collision Avoidance, 34, 65

Computer Programming, 8, 30

Computer Vision, 22, 224, 228

Conveyor Belt, 41, 46

**D**

Denavit-hartenberg Matrix, 32-33

Developmental Robotics, 1, 4-7, 38

Distance Sensor, 154

Domestic Robot, 12, 44, 81

**E**

Elastic Nanotubes, 124, 133

Electroactive Polymers, 47, 123, 133

Electromagnetic Spectrum, 58, 75

End Effectors, 54, 123-124, 141-144, 183-184

**F**

Force Torque Sensor, 157-158

Forward Kinematics, 31, 34, 116, 190

**H**

Hexapod Robot, 67, 70

Hexapoda Locomotion, 70

Human-robot Interaction, 6, 12, 125, 202, 224-225, 227

Humanoid Robots, 31, 35, 39, 41, 72-76, 78, 83, 133, 148, 179-180, 226

Hybrid Systems, 128, 195

**I**

Information Processing, 1, 11

Inverse Kinematics, 31, 33-34, 91, 104, 116, 118, 137

Inverse Velocity Kinematics, 108, 112-113

**J**

Jacobian Matrix, 100-102, 104-106, 108, 110-113, 115-116, 119

**K**

Kinematic Structure, 91-92, 94-96, 99-100, 103-105, 108, 111, 115, 206

**L**

Laboratory Robotics, 1, 20-22

Laws Of Robotics, 1, 25, 44, 48

Leg Configuration, 162

Legged Robots, 3, 39, 65, 67-70, 163-168, 170, 175, 177, 191

Linear Actuators, 89, 132-133

Load Balancing, 13, 25

**M**

Mars Rover, 3, 211

Microprocessor, 7, 11, 65

Microtiter Plate, 22

Military Robots, 44, 60, 62-64

Mining Robots, 44

Mobile Robots, 4, 12, 31, 36-37, 39, 41, 60, 64-66, 137, 148, 154, 198, 224

Motion Planning, 11, 33, 71, 145, 148, 188, 190-191, 226

Motor Control, 7, 14-16, 70

**N**

Nanorobots, 46, 230-231

Neurorobots, 14-15, 17-19

## O

Open-ended Learning, 4, 7
Open-source Robotics, 19

## P

Piezo Motors, 123, 133
Printed Circuit Boards, 41, 83
Proprioceptive Sensors, 58, 74

## R

Radio Control Servos, 129
Reconfigurable Robots, 47, 195
Rehabilitation Robots, 77, 213-214
Robot Kinematics, 30, 84, 96, 117
Robotic Arm, 21, 23, 31, 40, 54, 83, 123, 138, 140-141, 143, 214
Roomba, 10, 40, 44, 80-81

## S

Sensory Feedback, 1, 15, 184
Servo Robots, 84
Sheep Shearing Robots, 53, 224
Simultaneous Localization And Mapping, 42, 228
Skid Steer Drive, 171-172

Solar Engine, 7
Sonar Sensor, 150
Stepper Motors, 75, 123
Swarm Robots, 39, 47

## T

Tactile Sensors, 74, 148-150, 159, 185
Telepresence Robots, 39, 77, 83, 200-201, 226
Telepresence System, 39
Tilt Sensors, 74, 154

## U

Unmanned Aerial Vehicle, 43, 49, 62
Unmanned Space Probe, 43, 49

## V

Vex, 10

## W

Wheeled Locomotion, 70, 160-161, 170-171, 174-175

## Z

Zero Moment Point, 76, 165
Zeroth Law, 26-27

www.ingramcontent.com/pod-product-compliance
Lightning Source LLC
Chambersburg PA
CBHW061254190326
41458CB00011B/3664